Fundamentals of PROCESSING & FOOD ENGINEERING

THE AUTHORS

Rinku Grover has obtained his B.Tech (Agricultural Engineering) and M.Tech (Processing and Food Engineering) from College of Agricultural Engineering and Technology, Chaudhary Charan Singh Haryana Agricultural University, Hisar-INDIA. M.Tech research work was on "Design and development of Bael fruit pulper". Rinku Grover has qualified ICAR PG- JRF-2018 and ICAR- PG- SRF 2020 and Ph.D entrance exam (2020) of CCSHAU, Hisar. He has published various Hindi articles, research and review papers in reputed journals, Book chapters and Authored Book. Presently, he is pursuing Ph.D in Agricultural Engineering with specialization in Processing and Food Engineering at College of Agricultural Engineering and Technology, Chaudhary Charan Singh Haryana Agricultural University, Hisar-INDIA.

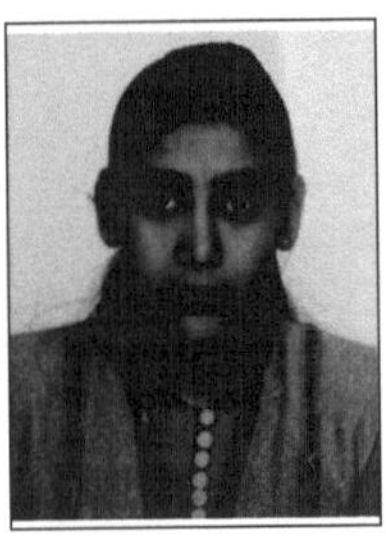

Dr. Raveena Kargwal has obtained his B.Tech (Agricultural Engineering) in 2014 and M.Tech (Processing and Food Engineering) from College of Agricultural Engineering and Technology, Chaudhary Charan Singh Haryana Agricultural University, Hisar in 2017. She has completed her Ph.D. (Processing and Food Engineering) from College of Agricultural Engineering and Technology, Chaudhary Charan Singh Haryana Agricultural University, Hisar (Haryana) in 2021. She has received Merit Scholarship during her Ph.D period from the CCS HAU, Hisar. Raveena has qualified Haryana LEET- 2011, HSTES, ICAR JRF-2015 and Ph.D entrance exam (2017) CCSHAU, Hisar. She has also qualified senior technical assistant exam held by CCS HAU, Hisar. She has worked as Senior Research Fellow (AICRP Project) in 2017-2018 in Processing and Food Engineering, CCS HAU, Hisar. At present, she is working as Senior Research Fellow in DST Project at Renewable and Bio Energy Engineering, CCS HAU, Hisar. She has more interest in Teaching and Research. She has published research papers in various NAAS rated international peer reviewed journals, book chapters, hindi articles. She has attended national / international conferences and also attended International workshop under SPARC project, MHRD, Govt. of India. She is life member of Indian Society of Agricultural Engineers (ISAE) since 2019.

Fundamentals of PROCESSING & FOOD ENGINEERING

(Useful for ICAR-JRF/SRF/ASRB-NET-ARS/CSIR-NET (LS)/BARC-OCES-Biosciences & Other Agricultural Competitive Examinations)

by

Rinku Grover

(Ph.D Scholar)

Department of Processing and Food Engineering

CCS Haryana Agricultural University, Hisar (India) 125001

Raveena Kargwal

(Ph.D)

Department of Processing and Food Engineering

CCS Haryana Agricultural University, Hisar (India) 125001

2025

Astral Cracker

A Division of

Astral International Pvt. Ltd.

New Delhi – 110 002

Cataloging in Publication Data--DK

Courtesy: D.K. Agencies (P) Ltd. <docinfo@dkagencies.com>

Grover, Rinku, author.

Fundamentals of processing & food engineering : useful for ICAR-JRF/SRF/ASRB-NET-ARS/CSIR-NET (LS)/BARC-OCES-biosciences & other agricultural competitive examinations / by Rinku Grover, Raveena Kargwal.

pages cm

Includes bibliographical references.

ISBN 978-93-5919-656-5

1. Food industry and trade. 2. Food industry and trade--Technological innovations. I. Kargwal, Raveena, author. II. Title. III. Title: Fundamentals of processing and food engineering.

LCC TP370.G76 2023 | DDC 664 23

Published by : **Astral Cracker**
A Division of
Astral International Pvt. Ltd.
– ISO 9001:2015 Certified Company
4736/23, Ansari Road, Darya Ganj
New Delhi-110 002
Ph. 011-43549197, 23278134
E-mail: info@astralint.com
Website: www.astralint.com

Dedicated

to

People's beloved President of India,

Dr. A.P.J. Abdul Kalam (Bharat Ratna)

"The best brains of the nation may be found on the last benches of the classroom"-Dr. A.P.J. Abdul Kalam

Preface

The authors are grateful for the inspiration and guidance of many senior and junior colleagues and friends through the course of preparation of this book. But most importantly, I will acknowledge that our efforts could be brought to this book form mainly due to the students. Recently, the authors contemplated to prepare a separate manuscript of objective questions on Processing and Food Engineering keeping in view the present-day need as a number of institutes/ colleges in the country are offering degree programs (bachelor/masters/ doctoral) in the disciplines of food technology, food engineering, agricultural processing, dairy engineering etc. The present books contains 13 chapters besides the questions on match the followings.

The chapters are on basic terminology, engineering properties, heat and mass transfer, material handling, post-harvest unit operation, storage engineering, processing technology and machinery, food packaging, dairy engineering, design of processing machinery, agricultural structures, agri-project planning and management and instrumentation and process control. Each chapter starts with brief one liner and multiple choices. Many of these questions have already been asked in various competitive exams. The authors are hopeful that the book will prove to be useful to the students pursuing their degree program in Agricultural Engineering, Food Technology/Engineering, Processing and Food Engineering, Dairy Technology/Engineering etc. The book is intended to prove useful for students appearing for competitive exams like JRF, SRF, NET, ARS, GATE etc. The support of our family members have also been very vital throughout the preparation of this manuscript. We wish the students/ readers all success in future endeavors. We will be highly obliged to get feedback for improving the book.

Rinku Grover

Raveena Kargwal

Contents

Chapter-1

Basic Terminology

Cleaning	Cleaning means the separation of undesirable and foreign matters from the desired grains/products.
Scalping	It refers removal of few large particles in initial process of cleaning.
Screening	Screening is a method of separating the grains/products into two or more fraction based on their size alone.
Ideal screen	A screen can be termed as ideal screen which separates the feed mixture in such a way that the largest particle of underflow is just smaller than screen opening, while the smallest particle of overflow is just larger than the screen opening.
Screen effectiveness	It may be defined as the ability of a screen in closely separating the feed into overflow and under flow according to its size.
Grading	The term grading refers to the separation of cleaned products into various quality fractions depending upon the various commercial values and other usages.
Sorting	The term sorting refers the separation of the cleaned products into various quality fractions based on size, shape, density, texture and colour.
Porosity	It is defined as percentage volume of inter-grain space to the total volume of grain in bulk. It depends on shape, dimensions and roughness of the grain surface.

Density	Density of a material is defined as mass per unit volume and is usually represented in the SI system by the unit's g/cm3 or kg/m3. The density of materials such as agricultural grains, which consist of many small particles, can be expressed in terms of solid density, particle density or bulk density. Solid density considers the mass and volume of the solid matter only and does not include any air spaces within the mixture. Particle density describes the mass per unit volume of an individual particle (e.g. corn kernel) from the sample. Bulk density, on the other hand, considers the total mass and total volume of a large quantity of the particles.
Apparent density	The density of the compressed mass or bulk is called the apparent density. The value of apparent density is higher by 20% than the normal bulk density taken into account in designing a silo.
Specific gravity	Specific gravity is a dimensionless term used to compare the density of different materials relative to water. Specific gravity of a substance is defined as the ratio of the substance's density to the density of water at the same temperature.
Sphericity	It is the ratio of diameter of a sphere of the same volume as that of the particle and the diameter of the smallest circumscribing sphere or generally the largest diameter of the particle.
Thermal conductivity	It is the rate of heat flow through unit thickness of material per unit area normal to the direction of heat flow and per unit time for unit temperature difference.
Specific heat	The amount of heat (kcal) that must be added to, or removed from 1 kg of a substance to change its temperature by 1°C.
Sensible Heat	The heat which cause an increase or decrease in the temperature of a body without changing its state.
Latent Heat	The latent heat does not cause a temperature rise within a body, but it affects a change in its state of being.
Total Heat	It is the sum of sensible heat and latent heat.
First law of thermodynamics	It states that, the matter can neither be created nor destroyed , however, it may be converted into different forms.
Second law of thermodynamics	It states that , the heat flow from higher temperature to lower temperature irrespective of material and mass of the bodies through which heat transfer takes place.
Drying	Drying is a thermo-physical and physico-chemical process that removes excess moisture from a product.
Drying time	It is the time at which one-half of the grain is fully exposed to the air entering the drying system.

Dehydration	It refers to the process of removing moisture from a product to the bone dry state.
Rehydration	It is the process of adding water to dried products in order to restore them to their original state.
Rehydration ratio	It is the weight of the drained rehydrated sample to the the weight of the dehydrated sample.
Sweating	It is the process of keeping dehydrated products in bins or boxes in order to equalise the moisture content.
Angle of repose	It is the angle formed by the base and slope of a cone formed by granular material falling vertically in free fall.
Angle of internal friction	The angle of internal friction is the tangent of the coefficient of friction.
Plane of rupture	The plane of rupture is the surface down which a wedge of material bounded by one wall face, the free surface, and the plane of rupture would slide if the bounding wall moved.
Shallow bin	A bin with relative dimensions such that the plane of rupture meet the grain surface before striking the opposite side wall.
Deep bin	A bin in which the plane of rupture collides with the opposite side wall before emerging from the grain.
Tripper	A belt tripper is a device for unloading that consists of two pulleys supported by a fixed or movable frame.
Effective belt tension	It is the tension created at the drive pulley as a result of transmitting the required power to move the loaded belt. It is the sum of the tensions to move the empty belt, the tension to move the load horizontally, and the tension to lift the material.
Milling recovery	Total milled rice obtained out of paddy; expressed as weight percentage of milled rice (including brokens) obtained from a sample of paddy.
Milling degree	It is a measure of the amount of bran removed from the brown rice.
Head rice	It refers to the milled whole rice of 6/8th and more of actual kernel size.
Broken rice	Rice kernels which are lesser than 6/8th of the actual size are called broken rice.
Total rice	It includes both broken rice and head rice.
Head rice recovery	Weight percentage of head rice (excluding brokens) obtained from a sample of paddy.
Conditioning	The conditioning is aimed to loosen the outer most hard layer of the grains known as husk to facilitate ease in its separation from the kernels for reducing the milling losses.

Extraction	It is the process of separating liquid from a liquid-solid system with the use of a solvent.
Expression	It is the process of separating liquids from solids by the applied pressure.
Filter press	It is a special equipment used to remove finely suspended particles in the juice.
Rice milling	It is the removal of husk and outer layer of bran from the paddy grain to get the acceptable white rice with the minimum broken and impurities.
Hulling/Dehusking	The purpose of hulling/dehusking is to remove the husk from paddy grain with minimum damage to the bran layer and as far as possible not to break the brown rice.
Glazing	It is the process of coating the head rice with a thin layer of talk and glucose for making the rice surface very shiny and more transparent.
Parboiling	It is a pre-milling treatment given to paddy prior to its milling to achieve maximum recovery of head rice and to minimise breakage.
Rough rice or Paddy rice	Rice as it comes from the field. Rice kernels are still encased in their inedible, protective hull.
Brown rice or husked rice	The least processed form of rice. It has the outer hull removed, but still retains the bran layer that gives it a characteristic tan colour and nut-like flavour, Brown rice is edible but chewier texture than white rice. Cooking time of brown rice is longer than milled rice.
White rice	White rice or rice after milling which includes removal of all or part of the bran and germ from the rough rice.
Conduction	The conduction is the mode of heat transfer in heating/cooling of opaque solids. Most metals are good conductor of heat. The free electrons provide an effective mechanism for carrying thermal energy from hotter to cooler portions of the metal. The conduction of heat can take place in a body only when different parts of the body are at different temperatures and the directions of heat flow is always from points of higher temperature to the points of lower temperature. An example of conduction is the heat transfer through the solid walls of a refrigerated store.
Convection	Convection heat transfer is the transfer of energy by the mass movement of groups of molecules.
Radiation	Radiation heat transfer is the transfer of heat energy by electromagnetic radiation. Radiation operates independently of the medium through which it occurs and depends upon the relative temperatures, geometric arrangements and surface structures of the materials that are emitting or absorbing heat.

Heat transfer	It is the movement of energy from one point to another by virtue of a difference in temperature. It is a dynamic process with temperature difference as a driving force.
Steady state	It is the state in which the temperature distribution remains constant with the time.
Unsteady state	It is the state when temperature distribution does not remain constant with time.
Coefficient of heat transfer (h)	It is defined as the heat transfer rate per unit area per unit temperature difference between a solid surface and fluid in contact with it.
Overall heat transfer coefficient (U)	It is the thermal conductance per unit area. It includes all heat transmission pro0cesses through the system.
Thermal conductivity (K)	It is the quantity or heat transmitted per unit time per unit area per unit temperature gradient.
Thermal diffusivity (α)	It determines the rate at which a non-uniform temperature distribution approaches equilibrium conditions.
Evaporation	Evaporation is a simultaneous heat and mass transfer process in which a portion of liquid is vaporised and the final product is still in the liquid form.
Concentrated juice	It is the juice obtained by the removal of a major part or their water content by vacuum evaporation or fractional freezing.
Condensation	The process of condensation is reverse of boiling. The condensation takes place when a saturated vapour comes in contact with the surface at lower temperature.
Distillation	Distillation is a separation process, separating components in a mixture by making use of the fact that some components vaporize more readily than others. When vapours are produced from a mixture, they contain the components of the original mixture, but in proportions which are determined by the relative volatilities of these components. The vapour is richer in some components, those that are more volatile, and so a separation occurs.
Active packaging	Incorporation or certain additives into packaging film or within packaging Containers to modify head space atmosphere and to extend product shelf life is referred as active packaging.
Modified Atmosphere Packaging (MAP)	It is defined as the packaging of a perishable product in an atmosphere that has been modified in such a way that its composition is different from that of air.
Packing	Its purpose is to protect the commodity by immobilizing and cushioning it.
Primary containers	Primary containers are defined as those that have direct contact with food.

Secondary containers	Containers that protect primary containers while being handled and transported.
Flexible films	The term "flexible film" is non-fibrous materials with a thickness of less than 0.25 mm. In general, they are heat sealable, suitable for high-speed filing, and add little weight to the product.
Shrink films	Shrink film packaging involves use of thermoplastic films that have been stretched or oriented during manufacturing and have property of shrinking with application of heat.
A-Pet films	These films are clear as glass with superior transparency and gloss. They have high mechanical properties and good chemical resistance. Excellent barrier properties, good printability, sealability and good bondability are other advantages of pet films.
Vacuum packaging	The earlier form of modified atmosphere packaging developed commercially and extensively for fruits, vegetables, primal cuts of fresh meats, cured meats, hard cheeses and ground coffee.
Saturated pressure	The water vapour in air obeys Daltan's law, therefore, it creates pressure independent of dry air. The water vapour pressure values of saturated water vapour are obtainable from steam table.
Humid heat	It is the specific heat of a mixture of 1 kg of dry air and Y kg of water vapour.
Humid volume	It is the volume of 1 kg of dry air and its vapour at given temperature and pressure.
Equilibrium moisture content (EMC)	It is the moisture content attained by a grain with respect to a set of atmospheric temperature and relative humidity.
Water activity	It is the ratio of the partial pressure of water in the food to the vapour pressure of water at the same temperature.
Dry-bulb temperature	This is the temperature of the air determined by a thermometer with a dry bulb.
Wet-bulb temperature	This is the temperature of the air measured by a thermometer with the bulb covered by a piece of wet cloth.
Dew-point temperature	This is the temperature at which the saturation state of a mixture of air and water vapour reaches during a cooling process at constant pressure
Absolute humidity	The ratio of kilogram of water vapour and kilogram of dry air is the absolute humidity.
Percentage humidity	This is the ratio of absolute air humidity with respect to the absolute air humidity at saturation point at same dry-bulb temperature.

Relative humidity	The ratio of actual vapour pressure in air-vapour mixture to the saturated water vapour pressure in air at same temperature is known as the relative humidity of air.
Equilibrium relative humidity	It is the percent relative humidity of an atmosphere in contact with a product at the equilibrium moisture content.
Refrigeration	Refrigeration is a process of lowering the temperature of a substance below that of its surrounding atmosphere. Refrigeration system is used to cool products, process or space at low, medium or very low temperatures.
Refrigeration cycle	Refrigeration cycle is comprised of four fundamental steps, viz, evaporation, compression, condensation and pressure reduction.
Chilling	Chilling of foods is a process by which their temperature is reduced to the desired holding temperature just above the freezing point of food, usually in the region of -2 to 2°C. The effect of chilling is only to slow down deterioration changes and the reactions as temperature dependent. So the time and temperature of holding the chilled food determine the storage life of the food.
One ton of refrigeration	It is defined as the quantity of heat required to be removed from one ton of water at 0°C within 24 hours to form one ton of ice.
Pasteurization of milk	It may be defined as heating every particles of milk to at least 61 °C and holding at that temperature for 30 minutes, or heating to at least 71°C and holding it at that temperature for at least 15 seconds. It destroys most of the bacteria in the milk and particularly all disease producing organisms. It also improves the keeping quality and taste of the milk.
Thermal death time	It is defined as the time needed for a particular fraction of the organisms to be killed.
Ozonization	It is a non-thermal method of killing micro-organisms using ozone through the oxidation of their cell membranes.
Sterilization	The object of sterilization is to destroy all microorganisms, that is, bacteria, veasts and moulds, in the food material to prevent decomposition of the food, which makes it unattractive or inedible. Also, sterilization prevents any pathogenic (disease-producing) organisms from surviving and being eaten with the food. Pathogenic toxins may be produced during storage of the food if certain organisms are still viable. Microorganisms are destroyed by heat, but the amount of heating required for the killing of different organisms varies. Also, many bacteria can exist in two forms, the vegetative or growing form and the spore or domant form. The spores are much harder to destroy by heat treatment than are the vegetative forms.

Pasteurization	Pasteurization is a heat treatment applied to foods, which is less drastic than sterilization, but which is sufficient to inactivate particular disease-producing organisms of importance in a specific foodstuff. Pasteurization inactivates most viable vegetative forms of microorganisms but not heat-resistant spores. Originally, pasteurization was evolved to inactivate bovine tuberculosis in milk. Numbers of viable organisms are reduced by ratios Order of 1015: 1. As well as the application to inactivate bacteria, pasteurization may be considered in relation to enzvmes present in the food, which can be inactivated by heat. The same general relationships as were discussed under sterilization apply to pasteurization. A combination of temperature and time must be used that is sufficient to inactivate the particular species of bacteria or enzvme under consideration. Fortunately, most of the pathogenic organisms, which can be transmitted from food to the person who eats it, are not very resistant to heat.
Parallel flow heat exchanger	It refers to two fluid streams flowing in the same direction through the heat exchanger.
Counter flow heat exchanger	It refers to two fluid streams flowing in the opposite directions through the heat exchanger.
Cross now heat exchanger	It refers to two fluid streams flowing in the perpendicular directions to the heat exchanger.
Fouling	It refers to the dirt and scale accumulated on the faces of the heat transfer surfaces. It reduces the heat transfer rate.
Regenerator	It is the heat exchanger in which the two fluid streams flow alternately through a matrix which exchange heat energy.
Cutting	Cutting is used to break down large pieces of food into smaller pieces suitable for further processing, such as in the preparation of meat for retail sales and in the preparation or processed meats and processed vegetables. In the process minimum deformation and rupture of the material results and the new surface created is more or less undamaged. An ideal cutting device is a knife of excellent sharpness and it should be as thin as practicable.
Grinding	In the grinding process, materials are reduced in size by fracturing them. Therefore, achieved by mechanical stress followed by upture and the energy required depends upon the hardness of the material and also upon the tendency of the material to crack-its friability.
Crushing	When an external force is applied on a material excess of its strength, the material fails because of its rupture in many directions.

Shearing	It is a process of size reduction which combines cutting and crushing.
Fineness modulus	It indicates the uniformity of grind in the resultant product. It can be determined by adding the weight fractions retained above each sieve and dividing the sum by 100.
Emulsions	Emulsions are stable suspensions of one liquid in another, the liquids being immiscible. Stability of the emulsion is obtained by dispersion of very fine droplets of one liquid, called the disperse phase, through the other liquid, which is called the continuous phase. The emulsion is stable when it can persist without change, for long periods of time, without the droplets of the disperse phase coalescing with each other, or rising or settling. The dispersed particles in the emulsion have a very large surface area, which is created in the process of emulsification.
Homogenization	Emulsions can be given increased stability by decreasing the size of the droplets either by impact or shearing the emulsion still further; the process is called homogenization.
Sedimentation	Sedimentation is very often used in the food industry for separating dirt and debris from incoming raw material, crystals from their mother liquor and dust or product particles from air streams.
Centrifugation	It is the separation of immiscible liquids or solids from liquids by the application of centrifugal force.
Filtration	It is the separation of solids from liquids by passing the mixture through a bed of porous material.
Floatation	In some cases, where it is not practicable to settle out fine particles, these can sometimes be floated to the surface by the use of air bubbles. This technique is known as flotation and it depends upon the relative tendency of air and water to adhere to the particle surface.
Membrane separation	Membrane separation involves partially separating a feed containing a mixture of two or more components by use of a semi-permeable barrier (the membrane) through which one or more of the species moves faster than another or other species.
Osmotic pressure	Water flows through the membrane from the dilute solution to the more concentrated one. The force producing this flow is called the osmotic pressure.
Washing	Washing is performed to remove the field dirt and microorganisms present of fruits and vegetables.

Waxing	Waxing of fruit is practiced mainly for cosmetic reasons because it improves appearance of the fruit. In addition to improve the appearance and shelf life of fruit, it slow down development of moulds and rind disorders such as stem end rind breakdown.
Viscoelasticity	It is a combined solid-like and liquid-like behaviour in which stress-strain relationship is time dependent.
Stress relaxation	Decay of stress with time when the material is suddenly deformed to a given deformation-constant strain.
Relaxation time	It is the rate of stress decay in a material subjected to a sudden strain. It is the time required for stress in the Maxwell model, approximately 37% of the original value.
Retardation time	It is the rate at which the retarded elastic deformation takes place in a material creeping under dead load. It is the time required for the Kelvin model, about 63% of its total deformation.
Creep	It is the deformation with time when the material is suddenly subjected to a dead load-constant stress.
Newtonian liquid	It is an ideal fluid in which the relationship between shear stress and shear rate is a straight line passing through the origin.
Non-Newtonian liquid	It is a liquid in which the relationship between the shear stress and shear rate is non-linear. If the flow curve is concave to the shear stress axis, the flow is called dilatent. if the flow curve is convex to the shear stress axis, the flow is called pseudoplastic.
Thixotropic fluids	These are fluids that show a decrease in shear stress with time of shear at a given shear rate.
Rheopectic fluids	These are fluids that show an increase in shear stress with time of shear at a given shear rate.
Coefficient of friction	It is the ratio between the force of friction and the force normal to the surface of contact.
Rheology	It is the science devoted to study of flow and deformation.
Deformation	Deformation or distortion is the relative displacement of points within a body. In general, it is accompanied by the change of volume or shape. The change of volume is caused by isotropic stress such as hydrostatic pressure. The change of shape is brought about by shear stresses. Material can be deformed by uniaxial compression, uniaxial tension, shear and bulk compression.
Bioyield point	It is the point at which increase in deformation occurs with a decrease or no change in force. This term is proposed for the biological materials.

Rupture point	It is a point at which the axially loaded specimen ruptures under a load. In biological materials, rupture may cause puncture of shell or skin. (Note. Bioyield point corresponds to a failure in microstructure while a rupture point coresponds to a failure in macrostructure of the specimen).
Ultimate strength	It is the stress corresponding to the rupture point.
Toughness	The work required to cause rupture in the material. This can be approximated by the area under the stress-strain or force-deformation curve up to the point selected as the rupture point.
Stiffness	Stiffness or rigidity is indicated by the slope of the initial straight line portion of the curve. The ratio of stress to strain in this more or less elastic region of the curve may be referred to as the modulus of elasticity or young's modulus.
Elasticity	The capacity of a material for taking elastic or recoverable deformation.
Plasticity	The capacity of a material for taking plastic or permanent deformation.
Poisson's ratio	The absolute value of the ratio of transverse strain to the corresponding axial strain resulting from uniformly distributed axial stress below the proportional limit of the material.
Proportional limit	The greatest stress at which a material is capable of sustaining without any deviation from proportionality of stress to strain (Hooke's law).
Resilience	It is the capacity of the material for storage of strain energy in the elastic range. The area under the unloading (stress-strain or force-deformation) curve is a meaure of resilience.
Mechanical hysteresis	The energy absorbed by a material in a cycle of loading and unloading evaluated as area between loading and unloading curve. It is an ability of the material to dissipate the strain energy as heat.
Rigidity	Rigidity or shear modulus is the ratio between shearing stress and the elastic shearing strain.
Gasification	It is a thermo-chemical reaction that converts solid carbon to combustible carbon monoxide in an air-tight chamber.
Gasifier	It is an equipment, in which gasification of biomass takes place.
Fermentation	It is the breakdown of complex molecules in an organic compound caused by a ferment such as yeast, bacteria, enzymes, and so on. It is a well-established and widely used technology for producing ethanol from grains and sugar crops.

Pyrolysis	It is the irreversible conversion of biomass into char, pyrolytic oils, and fuel gases via heating in an oxygen-free or low-oxygen environment.
Retention time	The time period during which feedstock are retained inside the digester is the retention time.

Chapter-2

Engineering Properties

ONE LINER:

- Kinetic friction force is less than static friction force.
- The density of water depends on temperature.
- The boiling point of water is lower than that of milk.
- The freezing point of water is higher than that of milk.
- The hydrophobic material has negative moisture binding energy.
- Thermal conductivity of air increased with increase in temperature.
- Stiffness is the index of rigidity of a material in elastic range.
- Low thermal conductivity materials are heated via dielectric heating.
- Water density is affected by temperature.
- As the temperature of evaporation lowers, the latent heat of vaporisation increases.
- The relative dielectric constant and the relative dielectric loss factor are important dielectric characteristics.
- Electric conductance is measured in mho/cm.
- Cream's specific gravity is always lower than milk's.
- Water has the highest specific heat value.
- The butter has an estimated fat level of 80%.
- Adding soluble substances to milk lowers the freezing point.
- If the acidity of milk is greater than 0.2 percent, it will deteriorate when heated or cooked.

- The specific heat of dry grains ranges from 0.35-0.45 kcal/kg °C.
- At the same temperature, the vapour pressure of the solution is lower than that of pure water.
- The vapor pressure of solution is less than that or pure water at the same temperature
- Latent heat of vaporization of bound water is determined by Othmer plot.
- The force required to break down the food tissues in chewing depends upon texture of the food.
- The stem end and apex of an oblate-shaped item are flattened.
- The sharpness of the solid's corners is measured by roundness.
- In a physical thing, shape and size are inextricably linked.
- When handling fruits and vegetables, the angle of repose is beneficial.
- Viscosity is a property of both gases and liquids.
- The majority of dairy products are poor electrical conductors.
- Farinograph is a device that evaluates the water absorption qualities of flour.
- Juice viscosity is proportional to temperature.

Questions

1. **Wheat grains having angle of repose is:**

(a) 30-45° (b) 30-40°

(c) 23-28° (d) 20-25°

2. **The angle of repose of paddy is:**

(a) 30-45° (b) 30-40°

(c) 23-28° (d) 20-25°

3. **In generalized Kelvin model of rheology first spring accounts for**

(a) Retarded strain (b) Instantaneous elastic strain

(c) Experienced flow in creep tests (d) Viscoelasticity

4. The maximum energy stored at

(a) Elastic limit (b) Plastic limit

(c) Limit of proportionality (d) None of these

5. Which the element spring represents elasticity in the mechanical model is replaced by........in electrical analog.

(a) Current (b) Resistance

(c) Inductance (d) Capacitance

6. The slope of the secant drawn from the origin to any point on the stress strain curve is called

(a) Initial modulus (b) Secant modulus

(c) Chord modulus (d) Tangent modulus

7. The ratio of transverse strain to the corresponding axial strain is as like

(a) Young's modulus (b) Shear Modulus

(c) Poisson's ratio (d) All the above

8. The strain that remains after the load that caused the deformation is referred to as:

(a) Set (b) Creep

(c) Shear Modulus (d) Young's modulus

9. The point on stress strain curve at which the stress is less than maximum attainable stress and increase in strain occurs without increase in stress is known as

(a) Rupture point (b) Yield point

(c) Elastic limit (d) None of these

10. The point on the stress-strain curve where axially loaded specimen ruptures under load is referred as

(a) Bio yield point (b) Rupture Point

(c) Elastic limit (d) None of these

11. The effect of many external forces working on a body is referred

(a) Stress (b) Strain

(c) Load (d) None of these

12. Load, which is considered to act a point

(a) Triangular (b) Uniformly distributed

(c) Point (d) None of these

13. The length of the body changes when a tensile or compressive force (P) operates on it.

(a) PL/AE (b) AE/PL

(c) PE/AL (d) PA/IE

14. The elongation of a circular tapered root is given by .... where P=Axial pull and d_1,d_2=diameter at two ends

(a) $4PL/\pi Ed1d2$ (b) $2PL/\pi Ed1d2$

(c) $4PL/\pi E(d1)^2(d2)^2$ (d) $2PL/\pi E(d1)^2(d2)^2$

15. Which of the following instrument is used to analyse flavoring components in food materials?

(a) Refractometer (b) Spectrophotometer

(c) Gas chromatography (d) None of theses

16. .... equation measures the sphericity of an objective?

(a) $(abc)^3/\sqrt{a}$ (b) $(abc)^{1/3}/a$

(c) $(abc)^2/\sqrt{a}$ (d) $(abe)^3$

17. .... shape is formed when an ellipse rotates about its major axis.

(a) Right circular cone/cylinder (b) Oblate

(c) Prolate (d) None of these

18. Which of the following the shape is formed when an ellipse rotates about its minor axis is:

(a) Oblate (b) Right circular cone

(c) Prolate (d) None of these

19. Volume of an irregular shape of a product can be measured by:

(a) Water displacement method (b) Thermister

(c) Dilatometer (d) All the above

20. For fruits and vegetables, volume of large objects can be calculated by

(a) Air comparison pycnometer (b) Platform scale method

(c) Specific gravity tube (d) None of these

21. Volume of smaller objects such as peas and beans can be determined by

(a) Bourdon tube (b) Platform scale

(c) Specific gravity balance (d) All the above

22. .... measured by specific gravity gradient tube

(a) Volume and weigh density (b) Vibrations

(c) Temperature (d) None of these

23. …………….. is not a physical property of agriculture material.

(a) Shape (b) Compressive strength

(c) Density (d) Size

24. ………….. shape is flattened at the stem and apex

(a) Oblate (b) Round

(c) Oblong (d) Concave

25. The angle of repose is generally friction for a grain of approximately the same moisture content than angle of internal

(a) Smaller (b) Higher

(c) No change (d) None of Theses

26. Artificial ageing brought about usually by (dry or wet) heat treatment of grain is called

(a) Conditioning (b) Steaming

(c) Curing (d) Parboiling

27. A device that assesses the dose of irradiation received by food in a qualitative or quantitative approach.

(a) Dosimeter (b) Tintometer

(c) Jelmeter (d) Salometer

28. Which of the following shape is tapered towards apex

(a) Conical (b) Oblate

(c) Oblong (d) Truncate

29. In which of the following shape, the horizontal cross section departs materially from circle.

(a) Regular (b) Irregular

(c) Square (d) Truncate

30. Air comparison pycnometer is used for determination of.

(a) True volume (b) Surface tension

(c) Surface area (d) Roundness

31. Pycnometer is used for …….

(a) Relative humidity (b) Specific gravity

(c) Surface area (d) Roundness

32. ……………solvent is used in pycnometer for measuring the seeds specific gravity.

(a) Ethyl alcohol (b) N-Hexane

(c) Toluene (d) Benzene

33. Sphericity represents..........characteristics of an agriculture produce.

(a) Shape (b) Size

(c) Cohesiveness (d) None of these

34. The ratio of largest projected area of object in the natural rest position to the area of the smallest circumscribing circle is called

(a) Size (b) Sphericity

(c) Roundness (d) Density

35. Specific heat expressed in

(a) Kcal/hr m °C (b) Kcal/Kg

(c) Kcal/Kg °C (d) Kcal/m °C

36. Surface area of the leaf can be measured by .... instrument.

(a) Planimeter (b) Pycnometer

(c) Day Apparatus (d) All the above

37. Rheology is the study of .... behavior

(a) Force, flow (b) Flow and Deformation

(c) Flow alone (d) None of these

38. .... parameters are used to express the mechanical behavior of material are

(a) Deformation, flow and time (b) Force, deformation and flow

(c) Force, stress and time (d) Force, deformation and time

39. Biological fluids mostly come under .... category.

(a) Non -Newtonian (b) Newtonian

(c) Plastic (d) Viscous

40. The surface of the pores in surface media exposed to fluid flow either per unit volume or per unit weight of the solid particles is referred as

(a) Specific surface (b) Surface area

(c) Porosity (d) Specific gravity

41. Geometric properties such as flakiness, graininess, lumpiness etc. denoted to

(a) Shape, size and orientation (b) Arrangements of constituents

(c) Appearance of Food (d)All of the above

42. The ideal elastic behavior is represented by

(a) St. Venant body (b) Newtonian body

(c) Hooken body (d) All the above

43. The ideal plastic behavior is represented by

(a) St. Venant body (b) Newtonian body

(c) Hooken body (d) All the above

44. The ideal viscous behavior is represented by

(a) St. Venant body (b) Newtonian fluid

(c) Hooken body (d) All the above

45. .... and .... comes under the broad rheological classification of flow.

(a) Viscosity and plasticity (b) Viscosity

(c) Plasticity (d) None of these

46. A small hole or wound on the surface of the fruits made by the pointed object or stem of other fruits is denoted by

(a) Split (b) Crack

(c) Puncture (d) Cut

47. One or more tortuous cracks radiating from point of impact is called

(a) Shatter cracks (b) Skin breaks

(c) Swell cracking (d) Spilt

48 The volume of prolate spheroid is expressed as if 'a' &'b is major and minor axis, followed by

(a) $4/3\ (\pi ab)^2$ (b) $4/3\ (\pi ab^3)$

(c) $4/3\ (\pi ab^2)$ (d) $4/3\ (\pi ab)^3$

49. Oblate volume spheroid is expressed as if 'a' & b' is major and minor axis, respectively as follows

(a) $4/3\ (\pi ab)^2$ (b) $4/3\ (\pi ab^3)$

(c) $4/3\ (\pi ba^2)$ (d) $4/3\ (\pi ab)^3$

50. Force divided area is known as

(a) Stress (b) Strain

(c) Viscosity (d) Surface tension

51. The stress calculated on the net cross section without considering geometric discontinuities such as hole, grooves is known as

(a) Shear stress (b) Normal stress

(c) Nominal stress (d) Torsional stress

52. .... law is used to calculate the terminal velocity of an object,

(a) Fourier's law (b) Wein's law

(c) Stokes law (d) Newton's law

53. Terminal velocity of particle is directly proportional to

(a) Diameter (b) Diameter square

(c) Square root of diameter (d) Diameter cube

54. ....properties are used for electrical sorting and grading of fruits and vegetables,

(a) Transmittance (b) Reflectance

(c) Shear stress (d) Both (a) and (b)

55. .... is the tendency of a liquid to flow.

(a) Fluidity (b) Consistency

(c) Viscosity (d) Plasticity

56. Angle of repose indicate the

(a) Ability to flow (b) Ability to resist

(c) Ability to break (d) None of these

57. .... has the same unit of stress

(a) Strain (b) Pressure

(c) Force (d) Surface Area

58. The porosity of wheat ranges

(a) 40-45% (b) 48-50%

(c) 50-55% (d) 65-70%

59. Paddy having porosity varies from

(a) 40-45% (b) 50-55%

(c) 48-50% (d) 65-70%

60. The tangent of angle of internal friction represents

(a) Angle of repose (b) Coefficient of friction

(c) Both (a) & (b) (d) None

61. The stress cause due to forces directed away from the plane on which they act is known as

(a) Compressive stress (b) Tensile stress

(c) Shear stress (d) Torsional stress

62. The stress component act tangentially to the plane on which the forces act is known as

(a) Compressive stress (b) Tensile stress

(c) Shear stress (d) Torsional stress

63. The shear stress on a transverse cross section caused by twisting is referred to as

(a) Shear stress (b) True Stress

(c) Fracture Stress (d) Torsional stress

64. The stress calculated on the basis of instantaneous cross sectional area instead of original cross sectional area is known as

(a) Normal Stress (b) Nominal stress

(c) True stress (d) None of these

65. The stress on minimum cross sectional area at the beginning of fracture is known as

(a) True Stress (b) Fracture Stress

(c) Torsional stress (d) Tensile stress

66. In stress -strain curve, the stress is proportional to the strain is called

(a) Proportional limit (b) Young's modulus

(c) Elastic limit (d) None of these

67. The relative displacement of points within a body is defined as

(a) Deformation (b) Strain

(c) Bio yield point (d) Rupture

68. Material capacity for taking deformation (plastic or permanent) is known as

(a) Viscosity (b) Elasticity

(c) Plasticity (d) None of these

69. Stress to corresponding strain within the proportional limit is known as

(a) Modulus of rigidity (b) Young's modulus

(c) Poisson's ratio (d) All the above

70. The slope of the stress-strain curve at the origin is known as

(a) Initial modulus (b) Secant modulus

(c) Chord modulus (d) Tangent modulus

71. The slope of the stress strain curve's initial straight line portion represents

(a) Stiffness or rigidity (b) Poisson's ratio

(c) Rupture Point (d) None of these

72. **The ratio of elastic deformation to the sum of elastic and plastic deformation when a material is loaded to a certain load and then unloaded to zero load is known as**

(a) Elasticity (b) Plasticity

(c) Degree of Elasticity (d) Degree of plasticity

73. **The stress corres ponding to the bio yield point is known as**

(a) Tensile strength (b) Ultimate Strength

(c) Bio yield strength (d) None

74. **The work required to cause rupture in material is known as**

(a) Toughness (b) Resilience

(c) Hysteresis (d) None of these

75. The area between the loading and unloading curve in stress -strain diagram is known as:

(a) Toughness (b) Resilience

(c) Hysteresis (d) None of these

76. **........................is a measure of the damping capacity or the ability of the material to dissipate the strain energy as heat**

(a) Toughness (b) Resilience

(c) Hysteresis (d) None of these

77. **The ratio of shearing stress to elastic shearing stain is referred as**

(a) Young's modulus (b) Shear modulus

(c) Hysteresis (d) None of these

78. **The stress that causes volume change is referred to as**

(a) Nominal stress (b) Isotropic stress

(c) Deviatoric stress (d) None of these

79. **A combined solid-like and liquid- like behavior in which the stress stain relationship is time dependent is known as**

(a) Viscoelasticity (b) Visco – Plasticity

(c) Rigidity (d) None of these

80. **The ratio of stress to strain is a function of time alone, not of the magnitude of the stress, in viscoelastic behavior,**

(a) Linear plasticity (b) Linear viscoelasticity

(c) Linear rigidity (d) All the above

81. When a material is suddenly deformed to a particular deformation, the stress decays with time.

(a) Stress relaxation (b) Relaxation time

(c) Creep (d) None of these

82. The rate of stress decay in a material subjected to a sudden strain is known as

(a) Retardation time (b) Relaxation time

(c) Creep (d) None of these

83. Relaxation time represents the stress behavior in

(a) Kelvin model (b) Maxwell model

(c) Generalized Maxwell model (d) None

84. The value of relaxation time in Maxwell model required for the stress to decay percent of its original value.

(a) 1/e or 37% (b) e

(c) e^2 (d) None of these

85. ....is more elastic

(a) Rubber (b) Plastic

(c) Brass (d) Steel

86. An ideal fluid in which the relationship between shear stress and rate is a straight line passing through the origin, the liquid is

(a) Newtonian liquid (b) Non- Newtonian liquid

(c) Real liquid (d) All the above

87. A liquid in which the relationship between shear stress and shear rate is non-linear is known as

(a) Newtonian liquid (b) Non- Newtonian liquid

(c) Real liquid (d) All the above

88. The flow curve of a non-Newtonian liquid is concave to the shear stress axis, the flow is referred

(a) Dilatant (b) Pseudo plastic

(c) Bingham (d) None of these

89. The flow of curve is convex to the shear axis in a non-Newtonian liquid, the flow is called

(a) Dilatant (b) Pseudo plastic

(c) Bingham (d) None of these

90. .... shows decrease in shear stress with time of shear at given shear rate is called

(a) Rheopectic fluids (b) Thixotropic fluids
(c) Newtonian fluids (d) All the above

91. Three fundamental properties of material which describes the rheological behavior are

(a) Elasticity, Plasticity, Viscosity (b) Deformation, Flow, Viscosity
(c) Elasticity, Time, Stress (d) Stress, Strain, Time

92. The stress strain relationship may depend on the rate of strain as well as higher time derivatives of the strain. This time dependency results in behavior called

(a) Viscoplastic (b) Viscoelastic
(c) Rigid (d) All the above

93. Generally, all agricultural products behave like

(a) Viscoelastic (b) Visco plastic
(c) Bingham (d) Rigid

94. Maxwell model connects

(a) Spring and dashpot parallel
(b) Dashpot and rubber in series
(c) Dashpot an spring in both parallel and series
(d) Spring and dashpot in series

95. Kelvin model connects

(a) Spring and dashpot parallel
(b) Dashpot and rubber in series
(c) Dashpot an spring in both parallel and series
(d) Spring and dashpot in series

96. Elements are coupled parallel in mechanical model, its equivalent in electrical network is

(a) Parallel (b) Series
(c) Both (a) & (b) (d) None of these

97. Elements are coupled series in mechanical model, its equivalent in electrical network is

(a) Parallel (b) Series
(c) Both (a) & (b) (d) None of these

98. Generalized Maxwell model usually represents

(a) Stress retardation (b) Stress relaxation

(c) Creep (d) None of these

99. Generalized kelvin model usually represents

(a) Stress retardation (b) Creep

(c) Stress relaxation (d) None of these

100. Product thermal diffusivity depended on

(a) Thermal conductivity (b)Density

(c) Specific heat (d) All the above

101. The range of sphericity of majority of grains is between.

(a) 1.32-2.32 (b) 0.32-1.00

(c) 1.32-2.00 (d) 0.32-1.32

102. .... constitutes all rheological and structural attributes of food product perceptible by means of mechanical, tactile, visual and auditory receptors?

(a) Flavor (b) Color

(c) Texture (d) Aroma

103. Dielectric constant of food material depends upon

(a) Temperature (b) Moisture content

(c) Density (d) Electrical conductivity

104. The angle of repose of maize is

(a) 30-45° (b) 30-40°

(c) 23-28° (d) 20-25°

105. The relation between bulk density, particle density and porosity is,

(a) $e = 1 - \frac{P.D}{B.D}$ (b) $e = 1 + \frac{P.D}{B.D}$

(c) $e = 1 + \frac{B.D}{P.D}$ (d) $e = 1 - \frac{B.D}{P.D}$

106. The porosity of corn varies from

(a) 40-45% (b) 50-55%

(c) 48-50% (d) 65-70%

107. The angle produced by the base of the cone and the slope of the cone formed by a free vertical fall of the grain mass to be a horizontal plane is,

(a) Angle of repose (b) Coefficient of friction

(c) Both (a) & (b) (d) None

108 Perishables products having moisture content (% w.b.)

(a) 60-70% (b) >80%

(c) 10-15% (d) None of these

109. The moisture content (% w.b) of semi perishable is

(a) 60-70% (b) >80%

(c) 10-15% (d) None of these

110. Rolling resistance is a .... property

(a) Physical (b) Rheological

(c) Frictional (d) Aerodynamic

111. The drag coefficient for an object of cone can be taken as

(a) 0.42 (b) 0.47

(c) 0.5 (d) 0.04

112. The young's modulus E, the shear modulus G and the poison's ratio are related by

(a) E=2G (1-μ) (b) E= 2G(1+2μ)

(c) E=2G(1+μ) (d) None of these

113. The Specific heat of dry grain is kcal/kg C

(a) 0.35-0.45 (b) 0.2-0.3

(c) 0.1-0.15 (d) 0.85-0.95

114. Angle of repose of very free flowing material is

(a) 25-30° (b) 30-38°

(c) 38-45° (d) >55°

115. Angle of repose of fair flowing material is

(a) 25-30° (b) 30-38°

(c) 38-45° (d) >55°

116. Angle of repose of cohesive material is_

(a) 25-30° (b) 30-38°

(c) 38-45° (d) 45-55°

117. The heat that causes a change of state without change in temperature is referred as,

(a) Sensible heat (b) Latent heat

(c) Specific heat (d) None of these

118. The latent heat of vanorization at 100°C & atmospheric pressure s about kJ/kg

(a) 2257 (B) 335

(c) 3230 (d) 3430

119. Viscosity of milk is .... centipoises

(a) 1 (b) 2

(c) 3.5 (d) 6

120. The triple point of water lies at

(a) 0.9 °C and 10mm of Hg pressure

(b) 0.01 °C and 4.59mm of Hg pressure

(c) 0.01°C and 10 mm of Hg pressure

(d) 0.9 °C and 4.59 mm of Hg pressure

121. The instrument used for measurement of force and deformation is

(a) Pentameter (b) Psychrometer

(c) Moisture meter (d) Instron

122. Paddy husk has a calorific content of about

(a) 3000 kcal/kg (b) 5600 kcal/kg

(c) 7000 kcal/kg (d) 4000 kcal/kg

123. Terminal velocity is an important aerodynamic property because of its application in

(a) Pneumatic conveying and separation of foreign materials

(b) Drying and storage

(c) Size reduction

(d) Mixing of ingredients

124. Specific heat of dry grains is about kcal /kg °C

(a) 0.1-0.15 (b) 0.35-0.45

(c) 0.6-0.75 (d) 0.9-1.2

125. Thermal conductivity of single grain is about .... kcal/m h°C

(a) 0.3-0.6 (b) 0.1-0.15

(c) 0.01-0.02 (d) None of these

126. Thermal conductivity of air is about .... kcal/m h°C

(a) 0.3-0.6 (b) 0.1-0.15

(c) 0.01-0.02 (d) None of these

127. On application of periodic auxiliary strain on a perfectly viscous liquid the shear stress is out of phase with applied strain by ....

(a) 180° (b) 90°

(c) 45° (d) 360°

128. Heating time of two materials of same thickness is changes

(a) Proportionality with thermal diffusivity

(b) Inversely proportion with thermal diffusivity

(c) Cannot be stated

(d) None of these

129. Thermal conductivity of ice .... times more than water

(a) 1 (b) 2

(c) 4 (d) 3

130. Dilatometer is used to measure

(a) Grain interporosity (b) Surface area

(c) True volume (d) Specific gravity

131. Electro static charges can be used to ....

(a) Measurement of moisture content

(b) Separation of grains

(c) Measurement of temperature

(d) All the above

132. The food quality such as taste, odor, colour etc., are measured in terms of

(a) Quick index (b) Organoleptic scale

(c) Score card (d) Hedonic scale

133. The property of material to form into thin sheet by beating is called

(a) Durability (b) Malleability

(c) Brittleness (d) Softness

134. End effect of viscometer is related to

(a) Capillary viscometer (b) Rotational viscometer

(c) Both (a) & (b) (d) Not related to viscosity

135. .... is not a method in the study of textural characteristics.

(a) Subjective (b) Objective

(c) Rheological method (d) Puncture test

136. Hardness of cereal grains decreases with the increase in

(a) Moisture content (b) Temperature

(c) Size (d) Volume

137. The quantity of wax required to coat on fruits is estimated based on

(a) Projected area (b) Surface area

(c) Roundness (d) All of the above

138. In a rotational viscometer relationship is established between

(a) Shear stress & shear rate (b) Density& shear rate

(c) Density& shear stress (d) Viscosity& density

139. Mechanical properties such as hardness, cohesiveness, viscosity, elasticity, adhesiveness and gumminess etc. dealing with ....

(a) Strength (b) Flow characteristics

(c) Both (a) & (b) (d) Nature of flow

140. The frictional force acting between the surface of contact at rest with respect to each other

(a) Static friction (b) Internal friction

(c) Kinetic friction (d) All the above

141. Load required to produce unit deflection is known as is called proof resilience

(a) Toughness (b) Stiffness

(c) Flexibility (d) None of these

142. Frictional coefficient of food grains depends on

(a) Grain shape (b) Surface characteristics

(c) Moisture content (d) All the above

143. Thermal conductivity of single grain is.................the bulk grain

(a) 3 to 4 times less than (b) 3 to 4 times greater than

(c) Equal to (d) 5 to 6 times greater than

144. The values of angle of repose for millet grains falls in the range of

(a) 15-20° (b) 30-35°

(c) 20-30° (d) 35-40°

145. Bulk density connects

(a) Mass and bulk volume (b) Mass and true volume

(c) Mass and volume (d) Bulk volume and porosity

146. **Firmness of fruits relates**
   (a) Force and depth of penetration (b) Force and time of application
   (c) Depth of penetration and time (d) All the above

147. **Advantages of dielectric heating are**
   (a) Uniformity of heating (b) High rate of temperature rise
   (c) Both (a) and (b) (d) None of these

148. **The frictional coefficient for food grains depends on**
   (a) Grain shape (b) Surface characteristics
   (c) Moisture content (d) All of the above

149. **Hardness of rice endosperm ranges from**
   (a) 6-9 kg/mm^2 (b) 9-11.5 kg/mm^2
   (c) 11.5-14.5 kg/mm^2 (d) 14.5 -18 kg/mm^2

150. **Terminal velocity of wheat varies from**
   (a) 6-9 m/s (b) 9-11.5 m/s
   (c) 11.5-14.5 m/s (d) 14.5-18 m/s

151. **Textural properties is measured by**
   (a) Colorimeter (b) Calipers
   (c) Polarimeter (d) Penetrometer

152. **Using Succulometer to measure texture is a .... method**
   (a) Nondestructive method (b) Destructive method
   (c) Sensory evaluation (d) None of these

153. **Quality control in food processing provides**
   (a) Customer satisfaction (b) Increased consumption and sales
   (c) Quality of product (d) All of the above

154. **Angle of repose is influenced by**
   (a) Size of the particle (b) Shape of the particle
   (c) Moisture content of the particle (d) All the above

155. **When the Hooken solid is under hydrostatic pressure, hydrostatic pressure(p)/volume strain (Ev) is given by**
   (a) Rigidity (b) Bulk modulus or incompressibilny
   (c) Viscoelasticity (d) Resilience

156. **Which of the following shows shear thickening behavior**
   (a) Pseudo plastic fluid (b) Rheopectic fluid
   (c) Dilatant fluid (d) Thixotropic fluid

157. A damage to plant tissue by external forces causing physical change in texture and lor eventually chemical alteration of colour, flavor and texture is known as

(a) Abrasion (b) Bruising

(c) Distortion (d) Crack

158. Cleavage without complete separation of the part the food is called

(a) Crack (b) Cut

(c) Shatter cracks (d) Skin break

159. Penetration or deviation by the sharp edge of an object is termed as

(a) Puncture (b) Cut

(c) Shatter eracks (d) Cracks

160. The moisture content (% w.b) of durable is

(a) 60-70% (b) >80%

(c) 10-15% (d) None of these

Answer Key

1.	(c)	21.	(c)	41.	(d)	61.	(b)	81.	(a)	101.	(b)	121.	(d)	141.	(b)
2.	(a)	22.	(a)	42.	(c)	62.	(b)	82.	(a)	102.	(c)	122.	(a)	142.	(d)
3.	(c)	23.	(b)	43.	(a)	63.	(d)	83.	(b)	103.	(b)	123.	(a)	143.	(b)
4.	(a)	24.	(a)	44.	(b)	64.	(c)	84.	(a)	104.	(b)	124.	(b)	144.	(c)
5.	(d)	25.	(b)	45.	(a)	65.	(b)	85.	(d)	105.	(d)	125.	(a)	145.	(a)
6.	(b)	26.	(c)	46.	(c)	66.	(a)	86.	(a)	106.	(a)	126.	(c)	146.	(a)
7.	(c)	27.	(a)	47.	(a)	67.	(a)	87.	(b)	107.	(a)	127.	(b)	147.	(c)
8.	(a)	28.	(a)	48.	(c)	68.	(c)	88.	(a)	108.	(b)	128.	(b)	148.	(d)
9.	(b)	29.	(b)	49.	(c)	69.	(b)	89.	(b)	109.	(a)	129.	(c)	149.	(c)
10.	(b)	30.	(a)	50.	(a)	70.	(a)	90.	(b)	110.	(c)	130.	(a)	150.	(b)
11.	(c)	31.	(b)	51.	(b)	71.	(a)	91.	(a)	111.	(c)	131.	(b)	151.	(d)
12.	(c)	32.	(c)	52.	(c)	72.	(c)	92.	(b)	112.	(c)	132.	(d)	152.	(b)
13.	(a)	33.	(a)	53.	(b)	73.	(c)	93.	(a)	113.	(a)	133.	(b)	153.	(d)
14.	(a)	34.	(c)	54.	(d)	74.	(a)	94.	(d)	114.	(a)	134.	(b)	154.	(d)
15.	(c)	35.	(c)	55.	(a)	75.	(c)	95	(a)	115.	(c)	135.	(c)	155.	(b)
16.	(b)	36.	(a)	56.	(a)	76.	(c)	96.	(b)	116.	(d)	136.	(a)	156.	(c)
17.	(c)	37.	(b)	57.	(b)	77.	(b)	97.	(a)	117.	(b)	137.	(b)	157.	(b)
18.	(a)	38.	(d)	58.	(c)	78.	(b)	98.	(b)	118.	(a)	138.	(a)	158.	(a)
19.	(a)	39.	(a)	59.	(c)	79.	(a)	99.	(c)	119.	(b)	139.	(c)	159.	(b)
20.	(b)	40.	(a)	60.	(b)	80.	(b)	100	(d)	120.	(b)	140.	(a)	160	(c)

Chapter-3

Heat and Mass Transfer

ONE LINER:

- In food processing operations, the transfer of heat energy is mostly achieved by heat exchanger.
- Heat transfer rates can be decreased by decreasing the area available for heat transfer.
- Heat transfer by convection follows Newton's laws.
- The use of plate heat exchanger is limited to low viscosity liquids.
- Heat exchanger used for transferring heat from fluid to another fluid.
- Emissivity is the ratio of emissive power of a body to the emissive power of a perfectly black body.
- According to second law of thermodynamics, heat transfer from higher to lower temperature takes place.
- The mean radius r for hollow cylinder of inner and outer radii for same radial heat flow rate is given by $[(r_2 - r_1/[\ln(r_2/r_1)]$.
- Mode of heat transfer takes place generally conduction, convection & radiation.
- The discipline of the heat transfer encompasses a great many fascinating area like design of steam generators, condensers etc
- Emissivity of any body is equal to its absorptivity at the same temperature is called Kirchoff's law.

- Frequency applied in industrial equipment in microwave heating is 2450 Hz.
- Thermal conductivity of ice is nearly 4 times the thermal conductivity of water.
- Fins are used in heat exchanger, to increase heat transfer by increase in surface area.
- Emissivity and absorptivity of surface are equal for radiation with equal temperature and wavelength
- As temperature increases, the wavelength of electromagnetic wave decrease
- Black body is body that absorbs and emits the maximum amount of energy at a given temperature.
- Radiation heat transfer between two surfaces can be reduced by introducing radiation shields between the surfaces.
- Biot number is a ratio relating convective and conductive heat transfer coefficients.
- In counter flow heat exchanger, the average temperature difference between the fluids is less as compared to parallel flow heat exchanger.
- Thermal conductivity of solid metals increase with rise in temperature.
- The material medium between the heat source and the receiver is not affected during the process of heat transmission by radiation.
- The heat flow per unit time across unit area when temperature changes in unit is called thermal conductivity.
- Thermal radiation occurs in the portion of electromagnetic spectrum between the wavelengths 0.1 to 100μm.
- When buoyancy forces are large, the flow occurs turbulent while viscous forces are large, the flow becomes laminar.
- The amount of heat required to convert a kg of grain moisture to water vapor is 650 Kcal.
- In cross flow heat exchangers, the direction of flow of one fluid is at right angle the other fluid.
- Fouling resistance is the additional resistance to heat transfer due to scale formation on surface of tubes of a heat exchanger.
- A 1-2 heat exchanger means single pass on shell side and double pass on tube side.

- ❒ Fouling refers to accumulation of material on the heating surface.
- ❒ Heat transfer rate in a heat exchange is directly proportional to LMTD.
- ❒ In heat exchangers, 25 % cut baffles are generally used.
- ❒ For calculation of film coefficient in heat exchangers, fluid properties should be calculated at average temperature.
- ❒ In free convection, film coefficient of air in horizontal pipes is proportional to 0.25 power of temperature difference.
- ❒ Out of still air, moving air, high viscosity liquids and condensing steam, surface heat transfer coefficient would be highest for condensing steam
- ❒ In natural (free) convection, the motion in fluid is caused due to difference in density resulting from temperature.
- ❒ Tao's charts can be used for computation of freezing time.
- ❒ Neumann problems relates to unit operation of drying

Questions

1. Emissivity (e) indicates the ability of

(a) Absorb incident radiation (b) Reflect incident radiation

(c) Transmit incident radiation (d) None of the above

2. If a surface absorbs incident radiation equally at all wavelengths, it is considered a

(a) Black body (b) Grey body

(c) Opaque body (d) Perfect absorber

3. An object which reflects all the incident radiation is called

(a) Grey object (b) White object

(c) Black object (d) None of above

4. .... factor does not affect the emissivity of a grey body?

(a) Temperature (b) Emission angle

(c) Wavelength (d) None of above

5. **The emissivity value for a grey body,.........at a particular wavelength with increase in temperature.**
   (a) Increases
   (b) decreases
   (c) Increases upto an extent and then decreases
   (d) remains constant

6. **Body has emissivity equal to its absorptivity at the same temperature, principle is known as**
   (a) Kirchoff's law (b) Planck's law
   (c) Boltzman's law (d) Maxwell's law

7. **The Stephan-Boltzmann law states that the energy flux emitted by a .... directly proportional to the fourth power of the absolute temperature.**
   (a) black (b) body is directly
   (c) opaque (d) any type of

8. **The value of Stefan Boltzmann constant (σ) is**
   (a) 22.6 W/m^2K (b) 22.6 W/m^2K^4
   (c) 5.6697x 10^{-8} W/m^2K^4 (d) 5.6697 x 10^{-8} W/m^2K

9. **Black body emits thermal radiation emitted at 150°C is**
   (a) 1612W (b) 1818 W
   (c) 1612 W/m^2 (d) 1818 W/m^2

10. **A polished metal has an emissivity of**
    (a) less than 0.1 (b) between 0.2-0.3
    (c) between 0.4-0.5 (d) more than 0.9

11. **At constant pressure, the heating of a gas is determined by**
    (a) Boyles law (b) Charles law
    (c) Gay lussacs law (d) Joules law

12. **Triple point temperature and pressure of water are**
    (a) 0.1 °C and 1 atm (b) 0.01 °C and 0.006028 atm
    (c) 0.1 °C and 0.06028 atm (d) 0 °C and 1 atm

13. **An isothermal process**
    (a) Internal energy decreases
    (b) Internal energy increases
    (c) There is no change in internal energy
    (d) Internal energy first increases and decreases

14. Diffusion is

(a) Mixing of gas in two containers at different pressure

(b) Mixing of two portions of fluid

(c) Mixing of two portions of fluids at different temperature

(d) Mixing of unlike fluids

15. Food can be cooked quicker in a pressure cooker because

(a) It is made of a metal

(b) Less heat lost

(c) Steam remains within the cooker

(d) None of these

16. The value of c_p/c_v for air is

(a) 1.2 (b) 1.4

(c) 1.8 (d) 2.3

17. The ....is an extensive property of a thermodynamic system?

(a) Pressure (b) Volume

(c) Density (d) Temperature

18. ....flow in which both fluids move in the same direction

(a) Parallel (b) Counter

(c) Cross (d) Newtonian

19. For a fluid at rest, pressure is the same

(a) Only at top surface (b) At all the points

(c) Only at bottom surface (d) None of above

20. Pressure of a refrigerant past the expansion valve in a refrigeration plant

(a) Decreases (b) Remains unchanged

(c) Increases (d) None of above

21. The product of Grashoff and Prandtl number is called

(a) Euler number (b) Cauchy Number

(c) Weber Number (d) Rayleigh number

22. Diffusion in solids follows

(a) Kick law (b) Fick law

(c) Kirchoff law (d) None of the above

23. **Salt is considered to be the**
   (a) Common refrigerant
   (b) Brominated refrigerant
   (c) First refrigerant
   (d) Both A and B both

24. **In natural convection, the heat transfer coefficient increases as the temperature increases.**
   (a) Biot number
   (b) Nusselt number
   (c) Grashof number
   (d) None of the above

25. **Emissivity is constant in**
   (a) White surface
   (b) Black surface
   (c) Gray surface
   (d) All the above

26. **Sun's radiation is mostly at ....wavelength.**
   (a) Long
   (b) Short
   (c) Both short and long
   (d) Medium

27. **The wave length range of thermal radiation is**
   (a) 1 to 4 μm
   (b) 0 to 1 μm
   (c) 1 to 100 μm
   (d) 0 to 4 μm

28. **Fin effectiveness is generally**
   (a) Less than one
   (b) Greater than one
   (c) Equal to one
   (d) None of the above

29. **The thermal conductivity of good conductors will ....as the temperature increases.**
   (a) Increase
   (b) Decrease
   (c) Initially increase and then decrease
   (d) Does not change

30. **As temperature increase, the thermal conductivity of insulators will**
   (a) Increase
   (b) Decrease
   (c) Initially increase and then decrease
   (d) Does not change

31. **Dry ice is called as**
   (a) Solidified water
   (b) Solified Freon 12
   (c) Solidified carbon-dioxide
   (d) Solidified Nitrogen

32. The "Adiabatic humidification" is an

(a) Isobaric process
(b) Isentropic process
(c) Isoenthalpic process
(d) Isothermal process

33. Fluid which is at rest, refers to

(a) Hydrostatic
(b) Hydrodynamics
(c) Fluid mechanics
(d) Both A and C

34. In a psychrometric chart the vertical scale shows the

(a) Dry bulb temp
(b) Wet bulb temp
(c) Absolute humidity in kg of water vapour per kg in dry air
(d) Pressure

35. When forced convection is compared to natural convection, the rate of heat transfer is

(a) Higher
(b) Lower
(c) Almost same
(d) Nothing in particular

36. Zeroth law of thermodynamics is for

(a) Pressure measurement
(b) Torque measurement
(c) Temperature measurement
(d) Power measurement

37. Manometer fluid should have

(a) High viscosity
(b) High surface tension
(c) High capillary effect
(d) None of the above

38. In adsorption isotherms, when water is bound very strongly to specific sites through an adsorption mechanics, this moisture is said as

(a) Bound moisture
(b) Free moisture
(c) EMC
(d) Monolayer moisture

39. One tone of refrigeration in terms of kcal/min is

(a) 30
(b) 50
(c) 3000
(d) 600

40. Vapour compression refrigerators are based on the principle that fluids absorb heat while changing from liquid phase to

(a) Solid phase
(b) Crystal phase
(c) Vapour phase
(d) Critical phase

41. **In vapour compression cycle, the refrigerant is in the form of dry saturated vapour before entering**

(a) Expansion valve (b) Condenser

(c) Compressor (d) Evaporator

42. **Nusselt number is**

(a) hl^2/k (b) hl/k

(c) hl/k^2 (d) h^2l/k

43. **Baffles in a shell and tube boiler serve the following purpose**

(a) To pass flue gases (b) To create more turbulence

(c) To decrease tube surface area (d) None of the above

44. **In natural convection process, heat transfer coefficient increase with increasing**

(a) Biot Number (b) Nusselt Number

(c) Grashoff Number (d) None of the above

45. **Pressure at a point is............in a static fluid.**

(a) Equal to the weight of the fluid above

(b) Equal in all direction

(c) Equal in all directions only if viscosity is zero

(d) Always directed upward

46. **Moisture in air is measured by**

(a) Thermister (b) Spectrophotometer

(c) Sling psychrometer (d) Orsat apparatus

47. **Nusselt number in convective heat transfer is equivalent to the convective mass transfer**

(a) Schmidt number (b) Sherwood number

(c) Prandtl number (d) Archemdes number

48. **The purpose of having more number of tubes in shell and tube exchanger is**

(a) To transfer heat

(b) To decrease the rate

(c) To increase surface area for higher heat transfer rate

(d) None of the above

49. Dry ice is called as

(a) Solidified water (b) Solified Freon 12

(c) Solidified carbon-dioxide (d) Solidified Nitrogen

50. Nusselt number in convective heat transfer is equivalent to following in convective mass transfer

(a) Schmidt number (b) Sherwood number

(c) Prandtl number (d) Archemdes number

51. ....is the primary refrigerant

(a) Sodium chloride (b) Calcium chloride

(c) Water (d) Ammonia

52. Bermoullis equation, as applied to a fluid flowing in a pipe, is based on the conservation of

(a) Energy (b) Mass

(c) Force (d) Momentum

53. Ratio of heat capacity rate for a heat exchanger is calculated by which of the following expressions

(a) Cmax/Cmin (b) Cmin/Cmax

(c) Cmax×Cmin (d) Cmax+Cmin

54. Bernoulli's law represents the following

(a) Conservation of mass (b) Conservation of energy

(c) Conservation of force (d) Conservation of momentum

55. Increasing the temperature of drying air will cause

(a) Increase its absolute humidity

(b) Decrease its absolute humidity

(c) Not change in its absolute humidity

(d) Change the absolute humidity marginally

56. The increase in the pressure of water will cause

(a) Increase the latent heat of vaporization

(b) No change in latent heat of vaporization

(c) Decrease latent heat of vaporization

(d) No change in boiling point

57. **At low velocity, the type of flow where layers of fluid seem to slide by one over the other without eddies being present is called**
(a) Turbulent flow (b) Laminar flow
(c) Transient flow (d) Fully developed flow

58. **Bernoullis law represents**
(a) Conservation of mass (b) Conservation of energy
(c) Conservation of force (d) Conservation of momentum

59. **Color code recommended for potable water in a plant is**
(a) Dark blue (b) Aqua
(c) Olive green (d) Violet

60. **The component velocity induced by a mixer in low viscosity liquids is**
(a) Longitudinal velocity (b) Rotational velocity
(c) Radial velocity (d) All of the above

61. **At constant volume, thermodynamic process taking place is termed as**
(a) Isobaric (b) Isochoric
(c) Isentropic (d) Adiabatic

62. **Convective heat transfer coefficient is expressed by the following dimensionless**
(a) Prandtl (b) Grashof
(c) Nusselt (d) Reynold

63. **A ton refrigeration is equivalent to**
(a) 50 kcal/min (b) 100 kcal/min
(c) 500 kcal/min (d) 1000 kcal/min

64. **Graphical representation of psychrometric chart is a**
(a) Physical properties of air
(b) Chemical properties of air
(c) Thermodynamic properties of air
(d) Physical properties of water vapor

65. **The part of a vapour compression refrigeration system, which absorbs heat from the refrigerated space is**
(a) Compressor (b) Condenser
(c) Evaporator (d) None of the above

66. Higher wet bulb depression will result in

(a) Longer drying time (b) Shorter drying time

(c) No effect on drying time (d) None of the above

67. The distance between the leading edge and the point at which the flow turns turbulent is

(a) Sub-critical length (b) Super-critical length

(c) Marginal length (d) Critical length

68. A refrigeration system can be used for

(a) Heating (b) Cooling

(c) Heating and cooling (d) Neither heating nor cooling

69. The heat from the surroundings is absorbed through .... in vapour refrigeration cycle

(a) Evaporator (b) Condenser

(c) Compressor (d) Fins

70. The dimensionless number in mass transfer operations which is analogues to Nusselt number in heat transfer is known as

(a) Lewis number (b) Sherwood number

(c) Peclet number (d) Schmidt number

71. The amount of moisture in a unit volume of air is referred to as

(a) Humidity (b) Absolute humidity

(c) Relative humidity (d) Specific humidity

72. In natural convection, the dimensionless number that connects buoyant and viscous forces is

(a) Nusselt number (b) Reynold number

(c) Grashoff number (d) Prandtl number

73. A mixture of air and water vapour is adiabatically cooled, the lowest temperature of the mixture thus achieved is equal to

(a) Dry bulb temperature (b) Wet bulb temperature

(c) Dew point temperature (d) Saturation temperature

74. The mass diffusivity to momentum diffusivity ratio is referred to as

(a) Reynolds number (b) Prandtl number

(c) Schmidt number (d) Grashoff number

75. The total radiant energy leaving a gray surface is known as

(a) Emissivity (b) Absorptivity

(c) Reflectivity (d) Radiosity

76. The emissivity of a gas body dependent on

(a) Pressure (b) Characteristic length

(c) Temperature (d) All of the above

77. For an opaque surface high reflectivity means .... Emissivity.

(a) High (b) Low

(c) Same (d) None of above

78. Insulating material have .... emissivity at angle near horizontal

(a) High (b) Low

(c) Same (d) Zero

79. Surface resistance is due to the .... of the surface

(a) Emissivity (b) Absorptivity

(c) Reflectivity (d) Transmitivity

80. In the case of, the flow direction has no influence on the performance.

(a) Condenser (b) Evaporators

(c) Dryers (d) Both A and B

81. Ideal fluid is

(a) Whose density is zero (b) Whose viscosity is zero

(c) Whose conductivity is zero (d) Whose velocity is Zero

82. Energy required in pumping fluid foods depends upon

(a) Thermal losses (b) Frictional losses

(c) Environmental conditions (d) None of the above

83. The continuity equation for an incompressible fluid has the following expression:

(a) $A_1v_1 = A_2v_2$ (b) $A_1\rho_1v_1 = A_2\rho_2v_2$

(c) $\rho_1v_1 = \rho_2v_2$ (d) None of the above

84. Potential flow is a flow of

(a) A compressible fluid with shear

(b) A compressible fluid with limited shear.

(c) An incompressible fluid with limited shear

(d) An incompressible fluid with no shear.

85. Creeping flow is fluid flow in which:

(a) Re < 1 (b) Re is between 300-400

(c) Re >10000 (d) None of the above

86. The hydraulic radius is defined as:

(a) Perimeter/area (b) Area/perimeter

(c) Area/perimeter2 (d) None of the above

87. If the diameter of a pipe is doubled, at same velocity of fluid, the discharge will increase by:

(a) 2 times (b) 4 times

(c) 8 times (d) None of the above

88. If fluid moves from a pipe of diameter D to another with diameter 2D, then the Reynold's number will:

(a) Four times (b) Double

(c) Half (d) Same

89. The distance between the leading edge and the line where the flow becomes turbulent is known as:

(a) Sub-critical length (b) Super critical length

(c) Marginal length (d) Critical length

90. If the flow velocity is increased, the boundary layer thickness for a given fluid and flow geometry will

(a) Increase (b) Decrease

(c) Remains constant (d) Increase exponentially.

91. Thermal buoyancy is to be studied in an enclosure with air as fluid. The relevant length scale is ....

(a) (v^2/g) (b) $(v^2/g)1/3$

(c) $(v_2/g)1/2$ (d) None of above

92. In conduction heat transfer, heat flux is proportional to the temperature gradient, i.e. q/A = -k, is called as

(a) Fourier's law of heat conduction (b) Kirchoff's law

(c) Pascal's law (d) None of the above

93. The relationship between thermal conductivity (k) and thermal diffusivity (α) are

(a) $\alpha= k\rho Cp$ (b) $\rho = \alpha kCp$

(c) $k= \alpha kCp$ (d) None of the above

94. ....material has higher thermal diffusivity than the others?

(a) Window glass (b) Wood

(c) Copper (d) Brick

95. As the viscosity in momentum transfer ...... is in heat transfer.

(a) Thermal conductivity (b) Thermal diffusivity

(c) Thermal resistance (d) None of above

96. The following is an analogy between heat, mass, and momentum transfer,

(a) Chilton- Colburn (b) Fick

(c) Newton (d) None of the above

97. Thermal coefficient of linear expansion is considered in case of ....

(a) Solids (b) Liquids

(c) Gases (d) Aall the above

98. The equation for heat transfer is written as, q =ΔT/(1/UA). In this equation, 1/UA is known as the

(a) Thermal potential (b) Resistance to heat transfer

(c) Overall heat transfer coefficient(d) None of the above

99. Raoults law is applicable to foods of following nature

(a) Ideal mixture (b) Non ideal mixture

(c) Non binary solutions (d) Emulsions

100. Log mean temperature difference in case of parallel flow compared to counter flow heat exchanger would be

(a) More

(b) Less

(c) Depend on type of heat exchanger

(d) Same

101. Diffusion of water molecules within solid food follows

(a) Kicks law (b) Ficks law

(c) Kirchoffs law (d) None of the above

102. A hollow sphere with a small hole behaves just like a perfect

(a) Grey body (b) White body

(c) Black body (d) Red body

103. Viscosity is a property that manifests.

(a) At fluid solid boundaries only

(b) Between two adjacent fluid layers in relative motion

(c) In uniform incompressible flows

(d) Only in turbulent flows

104. The mass of water vapour divided by the mass of dry air,

(a) Partial pressure of water vapor (b) Relative humidity

(c) Absolute humidity (d) Water ratio

105. The thermal resistance in case of sphere with inner and outer radius as r1 and r2 is given as

(a) Rsphere= $(r_2-r_1)/4\pi k r_2 r_1$ (b) Rsphere= $(r_2 r_1)/4\pi k(r_2-r_1)$

(c) Rsphere = $4\pi k\ (r_2-r_1)/\ r_2 r_1$ (d) Rsphere = $4\pi k r_2 r_1/(r_2-r_1)$

106. The quantity of heat required to convert one kg of ice at 0°C into vapour at 100°C is approx.

(a) 2258 kJ (b) 2676 KJ

(c) 3000 kJ (d) 3500 kJ

107. Some hot water is added to three times its mass of cold water at 10° C. The resulting temperature is found to be 20° C. Find the initial temp of hot water

(a) 45°C (b) 50°C

(c) 52°C (d) 56°C

108. When you add salt to ice, it.......

(a) Slow down the melting rate of ice

(b) Increase the melting rate of ice

(c) Not affect the melting rate of ice

(d) All of above

109. Which of the following factor has no effect on the thermal conductivity of a food?

(a) Temperature and pressure of surroundings

(b) Amount of air trapped between the cells

(c) Moisture content of food

(d) None of the above

110. In a heat transfer problem involving buoyancy effects, the relevant dimensionless number is

(a) Reynold's number (b) Nusselt's number

(c) Fourier number (d) Grashoff number

111. Fluid flow in circular pipe is said to be laminar if the Reynolds number is

(a) < 1 (b) 1

(c) < 2100 (d) > 10, 000

112. The normal value of heat transfer coefficient for air with natural convection is

(a) 5-25 W/m^2K
(b) 10-100 W/m^2K
(c) 500-10000 W/m^2K
(d) None of the above

113. The radiation exchange between two bodies depend on

(a) Temperature of the two bodies
(b) Angle factor between two bodies.
(c) Color of the bodies
(d) All the above

114. The angle factor between two bodies used in calculation of radiation heat exchange is also known as

(a) Correction factor
(b) Boltzman's coefficient
(c) Shape factor
(d) None of the above

115. The sum of all view facto from a given surface is

(a) Constant
(b) Varies with the shape alignment of surfaces
(c) Varies with the adiosity of the surface
(d) 1

116. The emissivity values of white paper and burnt toast are approx. ....and......, respectively.

(a) 0.1, 0.9
(b) 0.9, 1.0
(c) 0.1, 0.5
(d) 0.9, 0.1

117. Gasses are

(a) Opaque for high temperature radiation
(b) Opaque for low temperature radiation
(c) Transparent at short wavelengths
(d) Transparent at long wavelengths

118. The required surface characteristic combination for a solar collector is,

(a) High emissivity and low absorptivity
(b) High emissivity and high reflectivity
(c) High reflectivity and high absorptivity
(d) Low emissivity and high absorptivity

119. The emissivity of a gas body with a particular composition is determined by:

(a) Shape and temperature

(b) Partial pressure and shape

(c) Partial pressure and temperature

(d) All of these

120. The emissivity of a material depends on its shape, partial pressure, and temperature.

(a) O_2 is higher than that of N_2
(b) N_2 is higher than that of O_2
(c) O_2 is higher than that of CO_2
(d) CO_2 is higher than that of O_2

121. Mass transfer at molecular level can be likened to heat............

(a) Convection
(b) Conduction
(c) Radiation
(d) All of above

122. Mass diffusivity is expressed in ....

(a) m^2/s
(b) m/s
(c) m/s^2
(d) W/m.K

123. Mass transfer coefficient (hm) can be expressed as,

(a) m^2/s
(b) m/s
(c) m/s^2
(d) W/m.K

124. Mass transfer is due to .... gradient

(a) Temperature
(b) Concentration
(c) Pressure
(d) Moisture

125. Fouling resistance is due to .... of the heating surface during operation

(a) Increase
(b) Decrease
(c) Conductance
(d) Deterioration

126. The heat capacity of fluid stream is the product of

(a) Mass flow rate and specific heat

(b) Specific heat and enthalpy

(c) Specific heat and thermal conductivity

(d) Specific heat and thermal diffusivity

127. Storage type heat exchanges is also known as

(a) Shell and tube
(b) Plate
(c) Regenerative
(d) Single shell

128. For a given amount of heat conducted, higher the value of thermal conductivity .... will be the temperature difference.

(a) Lower (b) Higher

(c) Same (d) None of the above

129. In hollow cylinder, the temperature profile along the radius will be.....

(a) Linear (b) Exponential

(c) Logarithmic (d) Polynomial

130. Thermal conductivity expressed in SI system,

(a) W/mK (b) W/m^2K

(c) K/W (d) K/W^2

131. Thermal resistance is expressed in SI system,

(a) W/mK (b) W/m^2K

(c) K/W (d) K/W^2

132. Convective heat transfer coefficient units in the SI system is

(a) W/mK (b) W/m^2K

(c) K/W (d) K/W^2

133. In laminar flow, the velocity at a location with respect to time ....

(a) Increase (b) Decrease

(c) Constant (d) Initially increase and then decrease

134. The average convection coefficient along the length will .... in laminar flow.

(a) Increase (b) Decrease

(c) Constant (d) Initially increase and then decrease

135. Hydraulic mean diameter is expressed as

(a) 4A/P (b) 4P/A

(c) A^2/P (d) $4P^2$/A

136. Rayleigh number is the product of

(a) Reynolds and Prandtl number (b) Prandtl and Nusselt number

(c) Grashof and Prandtl number (d) Reynolds and Nusselt number

137. Graetz number is the product of

(a) Prandtl, Nusselt number and D/L

(b) Grashof, Prandtl number and D/L

(c) Reynolds, Nusselt number and D/L

(d) Reynolds, Prandtl number and D/L

138. Peclet number is the product of

(a) Reynolds and Prandtl number (b) Prandtl and Nusselt number

(c) Grashof and Prandtl number (d) Reynolds and Nusselt number

139. In film boiling main mode of heat transfer is

(a) Conduction (b) Convection

(c) Radiation (d) Convection and Radiation

140. In terms of mass flow G per unit width, the condensation Reynolds number is defined as

(a) $\mu/4G$ (b) $4G/\mu$

(c) $4G/\mu^2$ (d) μ/G

141. During the life of heat exchanger the performance will deteriorate due to:

(a) Regeneration

(b) Fouling

(c) Log Mean Temperature Difference

(d) Condensation of fluid on tube surfaces

142. The amount of water contained per unit mass of dry air is referred to as:

(a) Absolute humidity (b) Partial humidity

(c) Relative humidity (d) Equilibrium humidity

143. Zeroth law of thermodynamics relates to one of the following:

(a) Pressure measurement (b) Torque measurement

(c) Temperature measurement (d) Power measurement

144. On adsorption isotherms of a food, when water is said to be bound very strongly to specific sites through an adsorption mechanics, this moisture is said as:

(a) Bound moisture (b) Free moisture

(c) EMC (d) Monolayer moisture content

145. When a positive quantity K is divided into five parts, the maximum value of their product is:

(a) 5K (b) $(K/5)^2$

(c) $(5K)^5$ (d) $5(K/5)$

Answer Key

1.	(a)	31.	(c)	61.	(b)	91.	(c)	121.	(b)
2.	(b)	32.	(c)	62.	(c)	92.	(a)	122.	(a)
3.	(b)	33.	(a)	63.	(a)	93.	(c)	123.	(b)
4.	(d)	34.	(c)	64.	(c)	94.	(c)	124.	(b)
5.	(d)	35.	(a)	65.	(c)	95.	(b)	125.	(d)
6.	(a)	36.	(c)	66.	(b)	96.	(a)	126.	(a)
7.	(a)	37.	(d)	67.	(d)	97.	(a)	127.	(c)
8.	(c)	38.	(d)	68.	(c)	98.	(b)	128.	(a)
9.	(d)	39.	(b)	69.	(a)	99.	(a)	129.	(c)
10.	(a)	40.	(c)	70.	(b)	100.	(b)	130.	(a)
11.	(b)	41.	(c)	71.	(b)	101.	(b)	131.	(c)
12.	(b)	42.	(b)	72.	(c)	102.	(c)	132.	(b)
13.	(c)	43.	(b)	73.	(b)	103	(b)	133.	(c)
14.	(d)	44.	(c)	74.	(c)	104.	(c)	134.	(b)
15.	(c)	45.	(a)	75.	(d)	105.	(a)	135.	(a)
16.	(b)	46.	(c)	76.	(d)	106.	(c)	136.	(c)
17.	(b)	47.	(b)	77.	(b)	107.	(b)	137.	(d)
18.	(a)	48.	(c)	78.	(b)	108.	(a)	138.	(a)
19.	(b)	49.	(c)	79.	(a)	109.	(d)	139.	(c)
20.	(a)	50.	(b)	80.	(d)	110.	(d)	140.	(b)
21.	(d)	51.	(b)	81.	(d)	111.	(c)	141.	(b)
22.	(b)	52.	(a)	82	(b)	112	(a)	142.	(a)
23.	(c)	53.	(b)	83	(a)	113	(d)	143.	(c)
24.	(c)	54.	(b)	84	(d)	114	(c)	144.	(d)
25.	(c)	55.	(c)	85	(a)	115	(d)	145.	(b)
26.	(b)	56.	(c)	86	(b)	116	(b)		
27.	(c)	57.	(b)	87	(b)	117	(c)		
28.	(b)	58.	(b)	88	(c)	118	(d)		
29.	(b)	59	(a)	89	(d)	119	(d)		
30.	(a)	60.	(a)	90	(b)	120	(d)		

Chapter-4

Material Handling and Transport

ONE LINER:

- A belt conveyor's trough angle must not exceed a particular value in order to maintain a balance between capacity and longevity.
- In comparison to conventional conveyors, pneumatic conveyors demand more power per unit of material handled.
- A convex surface in the centre of the pulley lowers the belt's lateral displacement.
- Hook-link chains are only capable of transmitting electricity in one way.
- Conveyors using chains are less costly than those with belts.
- A belt conveyor can be used up to a maximum of 15 inclinations to horizontal.
- A pneumatic conveying system is less expensive.
- For grain processing, scraper conveyors are preferable.
- The material can be manually loaded onto the belt or by a machine that ensures a constant flow.
- A pneumatic conveyor can transport a broad range of goods, but it has a high power need.
- A screw conveyor may be utilised to incline up to 20 degrees from horizontal.
- A screw conveyor is ideal for transporting wet, sticky, or extremely abrasive goods.

- The pitch is equal to the diameter of screw for free-flowing materials, in screw conveyor.
- In a belt conveyor, an anti-friction bearing is employed.
- The top pulley of a bucket elevator is the driving pulley, while the lower pulley is the return and tension pulley.
- The speed of the belt in a bucket elevator is determined by the speed of the head pulley.
- The belt width in a bucket elevator should be 25 mm wider than the bucket width.
- The goods are lifted from the beginning to the end of the bucket elevator without any intermediate loading.
- For slow-moving materials, the screw pitch is generally 0.8 times the screw diameter in screw conveyors.
- The grain carrying capacity of a high capacity belt conveyor is less than 2000 tonnes per hour.
- Idlers have a significant role in belt conveyor efficiency.
- Bucket elevators have capacities ranging from 2 to 1000 tonnes per hour.
- The material handling capacity of a screw conveyor is proportional to the screw shaft's rotational speed.
- Conveyors have high mechanical efficiency.

Questions

1. **A set of operations that can be performed either by hand or by mechanical means or device to transport material and decrease drudgery is referred to as:**

(a) Filtration (b) Material separation

(c) Bulk storage (d) Material handling

2. **Material handling is concerned with:**

(a) Movement (b) Time and place

(c) Quantity and space (d) All of the above

3. **Conveyors are primarily used for .... movements**
   (a) Horizontal (b) Vertical
   (c) Inclined (d) Tangent
4. **.................. is not a material handling device**
   (a) Belt conveyor (b) Screw conveyor
   (c) Bucket elevator (d) None of the above
5. **......... is used for transporting the large quantities of materials over a very large distance at a low cost**
   (a) Chain conveyor (b) Screw conveyor
   (c) Belt conveyor (d) Pneumatic conveyor
6. **Belt conveyor of high capacity can be conveyed the grain up to the distance of**
   (a) 50 m (b) 100 nm
   (c) 120 mn (d) 200 m
7. **Inclined belt conveyor has maximum feed on angle of 10° but conveying angle should be**
   (a) 10° (b) 12°
   (c) 14° (d) 15°
8. **...........conveyor is an endless belt operating between two pulleys with its load supported on idlers**
   (a) Belt conveyor (b) Screw conveyor
   (c) Bucket elevator (d) All the above
9. **The idler spacing on the return side of the belt conveyor is kept at**
   (a) Same as the idler spacing on the load side
   (b) Half of the idler spacing on the load side
   (c) Two times the idler spacing on the load side
   (d) Four times the idler spacing on the load side
10. **The flat belt in belt conveyor used to transport**
    (a) Bulk grains (b) Bagged material
    (c) Fluid foods (d) None of these
11. **Which of the following shape of belt used to transport bulk grains in belt conveyoor**
    (a) Flat (b) V- shape
    (c) L- shape (d) All of these

12. **..................has the maximum mechanical efficiency**

(a) Belt conveyor (b) Chain conveyor

(c) Screw conveyor (d) Apron conveyor

13. **The belt width and troughing angle for high capacity belt conveyor is**

(a) 1500 mm and 45° (b) 1000 mm and 45°

(c) 1500 mm and 20° (d) 1000 mm and 45°

14. **The empty milk cans and plastic crates in a dairy plant are conveyed by**

(a) Belt conveyor (b) Chain conveyor

(c) Apron conveyor (d) Magnetic conveyor

15. **.......................is not a component of belt conveyor**

(a) Idlers (b) Blowers

(c) Pulley (d) Both A&C

16. **Belt conveyor have bearing**

(a) Frictional (b) Frictional and non-corrosive

(c) Anti-frictional (d) All the above

17. **Which statement is correct corresponding to belt conveyor**

(a) Relative motion exist between belt & product

(b) Blower is used in belt conveyor for conveyed grains

(c) No relative motion exist between belt and product

(d) All are correct

18. **Which of the following is the limitation for belt conveyor**

(a) Belt conveyor cannot be used for transport of grains

(b) Belt conveyor cannot be used for transport materials to elevated distances

(c) Belt conveyor cannot be used for transport of bagged materials

(d) All the above

19. **For transportation of grains in belt conveyor, the belt speed should not more then**

(a) 3.5 m/s (b) 2.5 m/s

(c) 4.5 m/s (d) 6 m/s

20. **The recommended belt speed for conveying grains in belt conveyor is .... m/s**

(a) 2.5-2.8 (b) 3.8-6.2

(c) 8-10 (d) 12-18

21. The selection of belt width in belt conveyor depends on

(a) Capacity (b) Speed

(c) Trough angle (d) All of these

22 Which of the following is the trough angle of belt conveyor to transport paddy grains

(a) 30° (b) 20°

(c) 40° (d) >50°

23. .... is the surcharge angle formed by paddy in belt conveying system

(a) 10° (b) 20°

(c) 35° (d) 45°

24. The efficiency of belt conveyor largely depends on

(a) Idlers (b) Constructional material

(c) Belt thickness (d) All of these

25. Which of the following idlers are suitable for transporting granular materials having angle of repose

(a) Flat belt idlers (b) Troughing idlers

(c) Cross belt idlers (d) Both (a) & (b)

26. The idler spacing in belt conveyor should not exceed

(a) 0.6m (b) 1.2m

(c) 2.4m (d) 1.8m

27. Generally, the idler spacing in belt conveyor is equal to

(a) Belt width (b) Belt thickness

(c) Belt length (d) None of these

28. The spacing of upper idlers in belt conveyor is .... the spacing of return idlers

(a) Equal (b) Less

(c) More (d) Cannot define

29. The main function of idler roller in belt conveyor is.

(a) To withstand the design work loads

(b) To protect the bearings

(c) Both (a) & (b)

(d) None of these

30. The spacing of troughing idlers in case of inclined belt operation is about the normal spacing

(a) Twice (b) Half

(c) Thrice (d) Equal

31. The grains in the belt conveying system can be delivered at any point along conveyor by using

(a) Tripper (b) Idler

(c) Buckets (d) None of these

32. The path of product flow while leaving the belt over the end pulley is,

(a) Circular (b) Ellipse

(c) Rectangular (d) Parabola

33. High capacity belt conveyors having a conveying capacity about tones

(a) 100 (b) 50

(c) 1000 (d) 2000

34. ....device is used for conveying of bulk material

(a) Belt conveyor (b) Screw conveyor

(c) Bucket elevator (d) All of these

35. The number of pulleys in bucket elevator is ....

(a) 1 (b) 2

(c) 3 (d) 4

36. The vertical lift of the bucket elevator is 36.

(A <50mn (b) 50-100m

(c) 100-150m (d) 150m

37. The conveying capacity of bucket elevator is about .... t/h

(a) 2-50 (b) 2-1000

(c) 1000-2500 (d) >2500

38. Which of the following is a type of bucket elevator

(a) Spaced bucket type (b) Continuous bucket type

(c)Both (a) & (b) (d) None

39 The spaced bucket centrifugal discharge type elevators are normally used for

(a) Powder (b) Grains

(c) Cement (d) None

40. Generally the pulley in bucket elevator is

(a) Driver pulley (b) Driven pulley

(c) Both (a)&(b) (d) None

41. The capacity of bucket elevator depends on

(a) Bucket size (b) Bucket spacing

(c) Conveying speed (d) All of these

42. Bucket elevator with belt carrier can be used to ..... speeds

(a) 1-2 m/s (b) 2.5-4 m/s

(c) 4-8.5 m/s (d) 10-12.5 m/s

43. The magnitude of centrifugal force in bucket elevator can be calculated by

(a) $C_f = \frac{wv^2}{gr}$ (b) $C_f = \frac{wv^3}{gr}$

(c) $C_f = \frac{wv^2}{g}$ (d) $C_f = \frac{vw^2}{gr}$

44. Velocity of bucket elevator can be calculated by

(a) $C_f = \frac{vw^2}{gr}$ (b) $n = 29.9/r$

(c) $n = 29.9/\sqrt{r}$ (d) $r/29.9$

45. The term back logging is used in ....

(a) Belt conveyor (b) Bucket elevator

(c) Screw conveyor (d) Pneumatic conveyor

46. The top angle of bucket in bucket elevator is

(a) 20-30° (b) 40°

(c) 60° (d) 80°

47. The bottom angle of bucket in bucket elevator is ....

(a) 20-30° (b) 40°

(c) 60° (d) 80°

48. In elevating of grains, the discharge from bucket elevator is a combination of

(a) Centrifugal (b) Gravitational

(c) Both (a) & (b) (d) Frictional

49. In a bucket elevator, if n is the rpm of the pulley and r is the radius of the pulley, then the pole length (l) is given by

(a) 1=895/n2 (b) 1-29.9/r

(c) 1-29.9/n (d) None of the above

50. The deep buckets, which are used for free flowing materials, have the front edge of the buckets sheared at an angle of

(a) 20° (b) 45°

(c) 65° (d) 90°

51. The screw conveyor consists of.....................shaped trough

(a) Tubular (b) U-shaped

(c) Spiral (d) All the above

52. Screw conveyor used for relatively

(a) Short (b) Long

(c) Both (a) & (b) (d) None of these

53. The pitch of standard screw is .... its diameter

(a) Equal (b) Half

(c) Twice (d) Thrice

54. The most common shape of troughs used in screw conveyor

(a) S-type (b) U-shaped

(c) L-shaped (d) All of these

55. .... conveyor is used to convey fine powder materials

(a) Chain conveyor (b) Screw conveyor

(c) Belt conveyor (d) Pneumatic conveyor

56. Screw conveyors are used for

(a) High capacities for long runs

(b) High capacities for short run

(c)Low and medium capacities for short run

(d) Low capacities for long runs

57. A screw conveyor inclined 15° will carry about .... % of the rated horizontal capacity

(a) 25 (b) 50

(c) 75 (d) 95

58. .... is normally used as a metering and mixing device at the outlet of the LSU dryer

(a) Belt conveyor (b) Screw conveyor

(c) Chain conveyor (d) Pneumatic conveyor

59. .... conveyor system is comparatively cheaper

(a) Belt conveyer (b) Chain conveyer

(c) Pneumatic conveyer (d) Screw conveyer

60. A screw conveyor inclined 25° will carry about .... % of the rated horizontal capacity

(a) 25 (b) 50

(c) 75 (d) 95

61. In ...., the granular materials move in a closed duct by a high velocity air stream

(a) Belt conveyor (b) Bucket elevator

(c) Pneumatic conveyor (d) All of these

62. .... is the part of bullock cart

(a) Spoke (b) Hub

(c) Yoke (d) All of the above

63. In .... the initial cost of the Conveyor of the conveyor is lower than any other conveyor with the same length and capacity

(a) Belt conveyor (b) Pneumatic conveyor

(c) Chain conveyor (d) Screw conveyor

64. Which type of conveyor moves granular material in closed duct by high speed current of air?

(a) Belt (b) Chain

(c) Screw (d) Pneumatic

65. The belt speed in belt conveyor should not increase.......... for transportation of grains,

(a) 2.0 m/s (b) 3.5 m/s

(c) 6.0 m/s (d) 7.0 m/s

66. Maximum inclination to carry food material with the belt conveyor is:

(a) 10° (b) 20°

(c) 30° (d) 40°

67. Material handling is concerned with

(a) Movement (b) Time & place

(c) Quantity & space (d) All of these

68 In bucket elevator, the material discharge chute is situated at

(a) Lower half but not bottom (b) Bottom

(c) Top (d) None of these

69. Pneumatic conveying of wheat is also termed as

(a) Two phase flow (b) Single phase flow

(c) Pipe flow (d) None of these

70. Which type of screw conveyor is commonly used for mixing of powder/ granular materials?

(a) Ribbon (b) Cut flight

(c) Double flight (d) Standard

71. Trolley conveyors consists of an overhead fastened together by chain track with trolleys that are

(a) T-beam (b) L-beam

(c) U-beam (d) I-beam

72. The conveyor used for transportation of sticky material is

(a) Apron conveyor (b) Belt conveyor

(c) Screw conveyor (d) Pneumatic conveyor

73. The gram of water vapour transmitted from 1 m^2 if film area in 24 hours when the vapour pressure difference of 1 mm of Hg is called

(a) Water vapour permeance

(b) Water vapour transmission rate (WVTR)

(c) Water vapour permeability

(d) Moisture transfer coefficient

74. The gram of water vapour transmitted from 1 m^2 if film area in 24 hours is known as

(a) Water vapour permeance

(b) Water vapour transmission rate(WVTR)

(c) Water vapour permeability

(d) Moisture transfer coefficient

75. What is the primary function of packaging material

(a) Containment (b) Protection

(c) Communication (d) convenience

76. The gram of water vapour transmitted from 1 m^2 if film area in 24 hours when the vapour pressure difference of 1 mm of Hg and 1 cm of film thickness is termed as

(a) Water vapour permeance (b) Water vapour transmission rate(WWVTR)

(c) Water vapour permeability (d) Moisture transfer coefficient

77. What are the roles of packaging materials (NET 2021)

(a) Protection (b) Communication

(c) Convenience and Containment (d) All of the above

78. Non crystalline of polymers are characterized by

(a) Glass transition temperature (b) Boiling temperature

(c) Melting temperature (d) All of the above

79. ....of belt conveyor is largely dependent on idlers?

(a) Length (b) Width

(c) Speed (d) Efficiency

80. Crystalline polymers are characterized by

(a) Boiling temperature (b) Glass transition temperature

(c) Both a and b (d) Crystalline melting temperature

81. In bucket elevator, the pulley width should be wider than the belt by:

(a) 10-20 mm (b) 50-60 mm

(c) 80-100 mm (d) 120-150 mm

82. In case of belt conveyor, the belt unit is coupled with the:

(a) Loading end of the belt

(b) Midway of loading and discharge end

(c) Discharge end of the belt

(d) None of the above

83. The capacity of screw conveyor is

(a) $\frac{\pi}{4}$(D-d) × p × n $\frac{\pi}{4}$(b) D-d) × p × n

(c) $\frac{\pi}{4}$(D-d) × p × n $\frac{\pi}{4}$(d) (D^2-d^2) × p × n

84. The capacity of a troughed belt conveyor is 40 m'/h. The cross-sectional area for 20 surcharge angle of 30.5 cm wide belt is 0.0072 square meter. The belt speed will be:

(a) 40 m/min (b) 80 m/min

(c) 93 m/min (d) 120 m/min

85. **Which of the following conveyor requires maximum power for the so amount of material movement**

(a) Belt conveyor (b) Screw conveyor

(c) Pneumatic conveyor (d) Chain conveyor

86. **Pneumatic conveving of wheat is also termed as:**

(a) Single phase flow (b) Two-phase flow

(c) Pipe flow (d) None of these

87. **In bucket elevator, top angle ....**

(a) 40° (b) 60°

(c) 80° (d) 90°

88. **Maximum inclination to carry food material with the belt conveyor is**

(a) 40° (b) 30°

(c) 20° (d) 10°

89. **In a pneumatic conveyer, during the discharge of bulk materials, air velocity in range of .... is required**

(a) 10-15 m/s (b) 15-30 m/s

(c) 30-40 m/s (d) 40-50 m/s

Answer key

1.	(d)	21.	(d)	41.	(d)	61.	(c)	81.	(c)
2.	(d)	22.	(b)	42.	(b)	62	(d)	82.	(c)
3.	(a)	23.	(b)	43.	(a)	63.	(b)	83.	(d)
4.	(d)	24.	(a)	44.	(c)	64.	(d)	84.	(c)
5.	(c)	25.	(a)	45.	(b)	65.	(b)	85.	(b)
6.	(d)	26.	(b)	46.	(d)	66.	(b)	86.	(b)
7.	(b)	27.	(a)	47.	(a)	67.	(d)	87.	(c)
8.	(a)	28.	(a)	48.	(c)	68.	(c)	88.	(c)
9.	(c)	29.	(c)	49.	(a)	69.	(a)	89.	(b)
10.	(b)	30.	(b)	50.	(c)	70.	(a)		
11.	(b)	31.	(a)	51.	(d)	71.	(d)		
12.	(a)	32.	(d)	52.	(a)	72.	(c)		
13.	(a)	33.	(d)	53.	(a)	73.	(a)		
14.	(b)	34.	(c)	54.	(b)	74.	(b)		
15.	(b)	35.	(b)	55.	(b)	75.	(b)		
16.	(c)	36.	(a)	56	(c)	76	(c)		
17.	(c)	37.	(b)	57	(c)	77	(d)		
18.	(b)	38.	(c)	58	(b)	78	(a)		
19.	(a)	39.	(b)	59	(c)	79	(d)		
20.	(a)	40.	(b)	60	(b)	80	(d)		

Chapter-5

Post-Harvest Unit operations

ONE LINER:

- Particle size is measured using standard screens that range from 76mm to 38μm.
- Temperature affects the density of water.
- Thermal conductivity of air increases with rising in temperature.
- The heat transfer between two places takes longer when a highly viscous product is thermally treated.
- Hydrophobic materials have negative moisture-binding energy.
- Milk's boiling point is higher than that of water.
- Stiffness is a measure of a material rigidity in the elastic range.
- Cream's specific gravity is always lower than milk's.
- The boiling point of milk is higher than the boiling point of water.
- Low thermal conductivity materials are heated via dielectric heating.
- As the temperature of evaporation decreases, the latent heat of vaporization increases.
- The relative dielectric constant and the relative dielectric loss factor are important dielectric characteristics.
- Electric conductance is measured in mho/cm.
- When Re 2, Stoke's law is used.
- Water has the highest specific heat value.

- Grain is separated by length using an indented cylinder.
- The aspirator and pneumatic separation both employ the grain's final velocity to separate it.
- Disc divider with a V' pocket Is utilised to separate materials that have a round form.
- The fat percentage of the butter is roughly 80%.
- The freezing point of milk drops when soluble substances are introduced.
- The angle of repose increases as the moisture content rises.
- Dry grains have a specific heat of 0.35-0.45 kcal/kg °C.
- The optimum moisture content for paddy harvesting is 20-22%
- The optimum moisture content for paddy storage is 12-14%
- The optimum moisture content for paddy milling is 13-14%
- Cyclone separator used for separation of particles from liquid as well as from gases.

Questions

1. Cleaned products are classified into various quality fractions depending upon commercial values

(a) Cleaning (b) Grading

(c) Sorting (d) Scalping

2. Which device is more widely used for separation and cleaning of seeds,

(a) Drier (b) Graders

(c) Screens (d) None of these

3. Screening is a method for separating seeds into two or more fractions based on their

(a) Texture (b) Size

(c) Shape (d) All of the above

4. **The process of separating cleaned product into several quality fractions based on shape, size, density, and colour,**
   (a) Sorting (b) Grading
   (c) Scalping (d) Cleaning

5. **Removal of large particles is referred to**
   (a) Grading (b) Scalping
   (c) Cleaning (d) Sorting

6. **A single screen is used to separate the grains into .... fractions**
   (a) One (b) Two
   (c) Three (d) Four

7. **The material which passes through a particular screen and retained over a subsequent screen are**
   (a) Undersize (b) Oversize
   (c) A or B (d) Intermediate particles

8. **The number of openings in Mesh are referred to**
   (a) Square inch (b) Linear inch
   (c) Square centimeter (d) None of these

9. **In the screen surface, minimum clear space between the edges of the opening is**
   (a) Aperture (b) Mesh
   (c) Open area (d) None of these

10. **The speed drops through the screen opening mostly due to**
    (a) Inertia (b) Gravity
    (c) Friction (d) None of these

11. **Commonly used ways of screen agitation is**
    (a) Revolving a cylindrical screen about a horizontal axis
    (b) By using stirrers
    (c) Revolving a cylindrical screen about a vertical axis
    (d) None of these

12. **Screens can satisfactorily clean and sort most of granular material when used along with**
    (a) Air blast (b) Thickness
    (c) Width (d) None of these

13. Which is not the purpose of screen motion

(a) To spread the material over the surface of the screen

(b) To cause the fine particle to settle at the subsurface

(c) To separate the foreign particles

(d) To discharge the oversize particles

14. The material which are retained over a screen surface are

(a) Undersize (b) Oversize

(c) Intermediate particles (d) None of these

15. The material which passes through the screen are

(a) Undersize (b) Oversize

(c) Intermediate particles (d) None of these

16. .... is mainly used for coarsest and rough separation.

(a) Grizzly (b) Cylinder sorter

(c) Shacking screen (d) All of these

17. The material flow path in grizzly is

(a) Passes through the bars

(b) Parallel to the length bars

(c) Perpendicular to the length bars

(d) None of these

18. Revolving screen is also called as

(a) Trammel (b) Grizzly

(c) Rotary screen (d) Both (a) and (c)

19. The vibration of the screen in vibratory screens is responsible for

(a) Providing passage to particle through the opening

(b) Continuous flow of particles along with screen

(c) It restricts screen clogging

(d) All of these

20. Essential shaft for counter balanced is present in

(a) Shacking screen (b) Gyratory screen

(c) Cylinder sorter (d) None of these

21. Screen's stroke length effects the

(a) Efficiency (b) Effectiveness

(c) Capacity (d) Both (a) and (b)

22. **The type of motion circular at the feed end and reciprocating and discharge end is presenting**

(a) Cylinder sorter (b) Gyratory screen

(c) Shacking screen (d) None of these

23. **Which type of screens are very popular for grinding, cleaning and separation**

(a) Trammel (b) Gyratory screen

(c) Vibratory screen (d) Circular screen

24. **In horizontal plane, which drive is used to convert horizontal reciprocating motion into rotary motion?**

(a) Eccentric drive (b) Rotary drive

(c) Combination of (a) and (b) (d) None of these

25. **Generally, screens are constructed by perforated**

(a) Sheet metal (b) Stainless steel

(c) Aluminum sheet (d) None of these

26. **In case of wire mesh ....is not a type of opening**

(a) Rectangle (b) Square

(c) Triangle (d) None of these

27. **Perforated sheet metal screens with round apertures are measured.**

(a) Opening area (b) Circumference

(c) Both (a) & (b) (d) None of these

28. **The direction of the oblong openings are kept in the grain flow over the screen direction of**

(a) Same (b) Diagonal

(c) Opposite (d) None of these

29. **The square opening in the wire mesh are measured by the number of opening per**

(a) mm (b) Sq. inch

(c) Inch (d) cm

30. **The ability of a screen to separate feed into overflow and underflow is referred as**

(a) Efficiency (b) effectiveness

(c) Both (a) & (b) (d) None of these

31. The operation of cleaning grading, separation of the product are formed by exploiting the difference in

(a) Electrical properties (b) Rheological properties

(c) Mechanical properties (d) Engineering properties

32. Perforated sheet metal screens with round apertures are measured by

(a) Circumference (b) Opening area

(c)Both (a) & (b) (d) None of these

33. Which material is used to prepare the screen cleaners

(a) Mild steel (b) Forged steel

(c) Stainless steel (d) None of these

34. The air screen cleaners does not use this type of cleaning system

(a) Scalping (b) Blowing

(c) Vibratory (d) Grading

35. At constant feed rate the bed thickness can .... by increasing the screen slope

(a) Reduced (b) Unchanged

(c) Increased (d) None of these

36. The higher percentage of open area increase

(a) Capacity (b) Efficiency

(c) Effectiveness (d) Both (a) and (b)

37. Screen opening size depends upon

(a) Size alone (b) Both size and shape

(c) Shape alone (d) None of these

38. In case of wheat, the optimum value of stroke length is

(a) 15cm (b) 17cm

(c) 15mm (d) 17mm

39. Which configuration used the series of decks attached with multiple decks,

(a) Inclined (b) Vertical

(c) Horizontal (d) None of these

40. At the bed thickness is .... by increasing the feed rate

(a) Increases (b) Remains same

(c) Decreases (d) Increased & then decreased

41. During the grading on the screen, Which type of pertorations used

(a) Small (b) Large

(c) Both (a) & (b) (d) None of these

42. Which type of perforations used for scalping on the screen

(a) Small (b) Large

(c) Both (a) & (b) (d) None of these

43. In air screen cleaner air is used for the removal of lighter particles based on the difference in the material

(a) Size (b) Shape

(c) Texture (d) Density

44. Which of the following accessories are used for improving the efficiency of the screen

(a) Screen bushes and rubber balls (b) Oil cloth cover

(c) a and b both (d) Eccentric unit

45. The percentage of open perforations is directly proportional to the

(a) Cleaning efficiency (b) Cleaning effectiveness

(c) Capacity (d) None of these

46. To keep screens clean .... is used, on the scalping Screens

(a) Tapper or hammer (b) Oil cloth cover

(c) Screen brushes (d) All the above

47. Disc separator separates the material on the basis of

(a) Differences in roundness

(b) Differences in length

(c) Differences in density or specific gravity

(d) Differences in shape

48. On the surface of in the disc separator, the present disc is known

(a) Indentations (b) Buckets

(c) Cups (d) None of these

49. In disc separator, which type of pockets used to separate the broken rice from the whole rice

(a) U-pocket (b) S-pocket

(c) V-pocket (d) R-pocket

50. The indents present on the inner surface of the indented cylinder separator are in shape

(a) Spherical (b) V- shape
(c) Hemi spherical (d) None of these

51. The triangle shaped perforated deck is present in

(a) Specific gravity separator (c) Intended cylinder separator
(b) Spiral separator (d) Disc separator

52. The grains are separated using a velvet roll separator.

(a) Shape (b) Surface texture
(c) Density (d) Both (a) & (b) both

53. Pneumatic separation working based on the difference in

(a) Mechanical properties (b) Rheological properties
(c) Electrical properties (d) Aerodynamic properties

54. Aspirator uses of the grain to separate into different fractions

(a) Density (b) Surface texture
(c) Terminal velocity (d) Specific gravity

55. The fan present in the pneumatic separator creates .... pressure than the atmospheric pressure

(a) Higher (b) Lower
(c) Maintain same pressure (d) None of these

56. The grains which are suspend in air velocity has terminal velocity

(a) Higher (b) Lower
(c) Maintain same pressure (d) None of these

57. Magnetic separator separates the grains based on

(a) Surface texture (b) Stickiness properties
(c) Both (a) & (b) (d) None of these

58. The particle is acted upon by ....forces, in the cyclone separator

(a) Gravitational (b) Centrifugal
(c) Weight of particles (d) Both (b) & (c)

59. The fluidized bed separator makes the separation based on the difference in

(a) Density (b) Size
(c) Terminal velocity (d) Both (a) & (b)

60. Which separator is used for the cleaning of lighter seeds such as radish, cabbage, lettuce, carrot etc.

(a) Fluidized bed separator (b) Pneumatic separator
(c) Velvet roll separator (d) All of these

61. Inclined belt draper separation takes place due to difference

(a) Shape (b) Surface texture
(c) Density (d) Both (a) & (b)

62. De-stoner separates on the basis of ....difference

(a) Mass (b) Texture
(c) Shape (d) Specific gravity

63. If the RH is below ...., natural air drying is recommended

(a) 60% (b) 70%
(c) 80% (d) 90%

64. Conduction type dryer is

(a) Tunnel dryer (b) Belt convection dryer
(c) Vacuum shelf dryer (d) None of the above

65. Direct contact dryer is

(a) Bin dryer (b) Rotary dryer
(c) Drum dryer (d) None of the above

66. Which is a moving bed dryer

(a) Cabinet dryer (b) Kiln dryer
(c) Rotary air dryer (d) None of the above

67. ....is a commercial grain dryer

(a) Bin dryer (b) Tunnel dryer
(c) LSU dryer (d)All of the above

68. In which of the dryers, vacuum can be maintained In improving the driving rate and product quality

(a) Tunnel dryer (b) Fluidized bed dryer
(c) Drum dryer (d) None of the above

69. Which dryer is best for peas

(a) Cabinet dryer (b) Fluidized bed dryer
(c) Pneumatic dryer (d) Tunnel dryer

70. For a short time for drying, which type of dryers utilize very high air temperatures of 500°C or more

(a) Spray dryer (b) Pneumatic dryer

(c) Rotary louver dryer (d) Flash dryers

71. Plug flow dryers are a specific type of

(a) Pneumatic dryer (b) Spray dryer

(c)Fluidized bed dryer (d) Tunnel dryer

72. .... dryer can act as a desolventizer

(a) Rotary dryer (b) Screw conveyor dryer

(c) Pneumatic dryer (d) None of the above

73. Accomplished fruit pulp drying by

(a) Fluidized bed dryer (b) Tray dryer

(c) Drum dryer (d) Spray dryer

73. ....gives the maximum spongy texture of food material

(a) Cabinet dryer (b) Fluidized bed dryer

(c) Spray dryer (d) Freeze dryer

74. ....is not used to dry paste materials

(a) Cabinet dryer (b) Drum dryer

(c) Bin dryer (d) None of the above

75. Freeze drying time is proportional to the

(a) Square of the thickness of material being drying

(b) Thickness of material being drying

(c) Cube of the thickness of material being drying

(d) Forth power of the thickness of material being drying

76. Rietz mill is able to grind materials below .... μ size

(a) 5 (b) 10

(c) 15 (d) 100

77. Grains with higher moisture content are more .... to grind than dry grains

(a) Easy (b) Difficult

(c) Cannot define (d) None of these

78. Fibrous materials require .... power than non-fibrous in grinding operation

(a) Less (b) More

(c) Same (d) All of the above

79. In convention drying of agricultural materials, rate of drying can be controlled by

(a) Air velocity (b) Air temperature

(c) Humidity (d) All the above

80. During dehydration of vegetables, the temperature of dehydration should be

(a) 40 to 50°C (b) 60 to 70°C

(c) 80-90°C (d) 100-115°C

81. Microwave oven is as dryer is an example of

(a) Vacuum drying (b) Di electric drying

(c) Freeze-drying (d) Radiation drying

82. LSU type dryer was developed in

(a) India (b) England

(c) USA (d) Canada

83. Which of the following dryer more suitable for cereal grain is

(a) Tunnel dryer (b) Rotary dryer

(c) Tray dryer (d) Drum dryer

84. Which device is not an oil expression

(a) Hydraulic press (b) Screw press

(c) Crusher (d) None of these

85. The edible portion of paddy is

(a) Germ (b) Bran

(c) Rice (d) Husk

86. The weight of hull is .... % of total weight paddy grain

(a) 4-6 (b) 18-22

(c) 70-72 (d) 90

87. The weight of endosperm is............... %of the total weight paddy grain

(a) 4-6 (b) 18-22

(c) 70-72 (d) 90

88. The moisture content should be ....% in paddy at the time of milling

(a) 9 to 10 (b) 11 to 12

(c) 13 to 14 (d) 16 to 18

89. The barrel in serew press is made of_

(a) Case hardened steel (b) Mild steel

(c) Cast iron (d) Copper material

90. The weight of bran is .... % of the total weight of paddy grain

(a) 4-6 (b) 18-22

(c) 70-72 (d) 90

91. Separation of oil from seeds by applying pressure is called

(a) Expression (b) Extraction

(c) Leaching (d) All of these

92. The process of de-waxing in oils is performed to remove

(a) Wax (b) Organic impurities

(c) Moisture (d) All the above

93. Palletization process is

(a) Densification (b) Gasification

(c) Oxidization (d) Liquefaction

94. Separation of oil from seeds by using solvent is called

(a) Expression (b) Extraction

(c) Leaching (d) All of these

95. Grinding of spices break down or expose the tissues with inevitale loss of

(a) Color principles (b) Aromatic principle

(c) Moisture principles (d) Weight principle

96. Groundnut decorticators rotate at about

(a) 90 rpm (b) 110 rpm

(c) 210 rpm (d) 510 rpm

97. Generally, the solvent in oil extraction is

(a) Methyl alcohol (b) Benzene

(c) n-Hexane (d) All of these

98. Gelatinization temperature for paddy is about

(a) 50°C (b) 60°C

(c) 70°C (d) 90°C

99. Parboiled rice takes .... cooking time than raw rice

(a) Less (b) Longer

(c) Equal (d) None of these

100. Soaking of paddy is .... process

(a) Diffusion (b) Heat transfer

(c) Ionic process (d) None of these

101. Parboiling rice is .... process

(a) Hydrothermal (b) Exothermal

(c) Endothermic (d) All of these

102. Which of the not a method of parboiling

(a) CFTRI method (b) Traditional method

(c) Jadavpur university method (d) None of these

103. A bran of raw rice contains .... oil than parboiled rice

(a) Less (b) More

(c) Equal (d) None of these

104. The deoiled cake coming out of the extraction unit has ....% solvent by weight

(a) 0-25 (b) 25-40

(c) 40-60 (d) 60-80

105. The milled whole rice of 6/8 and more of actual kernel size in rice milling process is called

(a) Total rice (b) Small broken

(c) Big Broken (d) Head rice

106. The milled whole rice of 4/8 and more of actual kernel size in rice milling process is called

(a) Head rice (b) Big broken

(c) Small broken (d) Total rice

107. The machine used for paddy hulling in India is

(a) Engle berg huller (b) Under runner disc huller

(c) Rubber roll Sheller (d) None of above

108. The effectiveness of the roller mill rely on

(a) Clearance between rolls

(b) Geometrical parameters of rolls

(c) Kinematic properties of rolls

(d) All of these

109. Cryogenic grinding is mainly used for

(a) Grains (b) Fruits

(c) Spices (d) All of these

110. Refining oil is the process of

(a) Neutralization of vitamins

(b) Concentration of free fatty acids

(c) Neutralization of free fatty acids

(d) Concentration of vitamins

111. The deoiled cake of solvent extracted soybean oil contains % of oil residual

(a) 0.1 (b) 0.5

(c) 1 (d) 5

112. Deodorization of oil is refining process is accomplished with

(a) Heating (b) NaOH

(c) Nacl (d) H_2SO_4

113. Protein content in brown rice is .... %

(a) 1-3 (b) 3-5

(c) 6-7 (d) >10

114. The air blast freezer is an example of a

(a) Quick freezer (b) Slow freezer

(c) Ultra-rapid freezer (d)Rapid freezer

115. The higher degree of gelatinization mainly reflects on

(a) Hardness (b) pH

(c) Sugar content (d) None of these

116. Parboiling increases the total yield by .... %.

(a) 1-1.5 (b) 1.5-3

(c) 3-5 (d) 5-8

117. The final eutectic temperature for ice cream is approximately

(a)-5 °C (b) 30 °C

(c)-55 °C (d) -88°C

118. Boiling point of liquid nitrogen

(a)-196 °C (b) -78.5 °C

(c)-29.8 °C (d) None of the above

119. .... is commercially used for preparation of ice cream

(a) Scraped surface freezer (b) Air blast freezer

(c) Chest freezer (d) Immersion freezer

120. The water molecules tend to form ground before the change of phase in the freezing process which is known as

(a) Supercoiling (b) Nucleation

(c) Planck's behavior (d) Thawing

121. The process used for pre-concentration of liquid foods prior to driving operation

(a) Freezing (b) Evaporation

(c) Dehydration (d) Hardening

122. Discolourisation occurred due to .... in rice during parboiling

(a) Caramelization (b) Mallard reactions

(c) Pyrolysis (d) None of these

123. The oil content obtained from bran of raw rice is about ....%.

(a) 10-20 (b) 25-30

(c) 35-52 (d) >80

124. The essential pre-milling operations of wheat are

(a) Separation of impurities (b) Tempering

(c) Conditioning (d) All of the above

125. The ratio of final volume of rice to initial volume of rice is called

(a) Swelling capacity (b) Expansion ratio

(c) Compression ratio (d) None of these

126. The hardness of rice endosperm is Kg/mm^2

(a) 6-9 (b) 11.5-14.5

(c) 15.6-21.2 (d) 21.5-30.5

127. ....evaporator is best suitable for high viscous foods

(a) Rising film evaporator (b) Horizontal tube evaporator

(c) Agitated film evaporator (d) All of these

128. Percentage germ in wheat kernel contains, approximately

(a) 10 (b) 14.5

(c) 6 (d) 2.5

129. The oil content of bran obtained from parboiled rice about ....%

(a) 10-20 (b) 25-30

(c) 35-52 (d) >80

130. The boiling point of a liquid food increases in evaporators because

(a) The concentration of solids increases with evaporation

(b) The evaporator chamber has higher pressure than the normal atmosphere

(c) Both (a) and (b)

(d) None of the above

131. A short tube evaporator is not suitable for

(a) Low viscous foods (b) High viscous foods

(c) Low pH foods (d) High pH food

132. The barometric leg is a device to

(a) Discharge condensed vapour from high pressure region to low pressure region

(b) Discharge condensed vapour from low pressure region to low pressure region

(c) Allow steam to enter to steam chest without pressure drop.

(d) Maintain vacuum in the evaporator

133. The term *Calendria' is used in

(a) Dryer (b) Freezer

(c) Evaporator (d) All of these

134. The basic objective of multiple-effect evaporation system

(a) To improve Product quality

(b) To improve Throughput capacity

(c) To improve Economy of operation

(d) None of the above

135. The flour extraction from wheat is ....

(a) 80-84% (b) 70-74%

(c) 65-70% (d) 60-64%

136. The term bulger represents

(a) Polished and cracked pulses (b) Polished and cracked wheat

(c) Polished and cracked paddy (d) Polished and cracked maize

137. Peeling of wheat during lye treatment is carried by using

(a) Manual power (b) Hydro cyclone

(c)H_2SO_4 (d) None of these

138. The time required for husking in rubber roller sheller is calculated by

(a) Bernoulli's equation (b) Henderson's equation

(c)Kupit's equation (d) All of these

139. The hardness of barley endosperm is Kg/mm^2

(a) 6-9 (b) 11.5-14.5

(c) 15.6-21.2 (d) 21.5-30.5

140. The grinding operation is effected by.................of grains

(a) Type & variety (b) Moisture content

(c) Mechanical properties (d) All of these

Answer Key

1.	(b)	21	(b)	41	(b)	61.	(d)	81.	(b)	101.	(a)	121.	(b)
2.	(c)	22	(b)	42	(a)	62.	(d)	82.	(c)	102.	(d)	122.	(b)
3.	(b)	23.	(c)	43.	(d)	63.	(b)	83.	(b)	103.	(c)	123.	(a)
4.	(a)	24.	(a)	44.	(c)	64.	(c)	84.	(c)	104.	(b)	124.	(d)
5.	(b)	25.	(a)	45.	(a)	65.	(a)	85.	(c)	105	(d)	125.	(a)
6.	(b)	26.	(c)	46.	(a)	66.	(c)	86.	(b)	106.	(b)	126.	(b)
7.	(d)	27.	(c)	47.	(b)	67.	(d)	87.	(c)	107.	(a)	127.	(c)
8.	(b)	28.	(a)	48.	(a)	68.	(c)	88.	(c)	108.	(d)	128.	(d)
9.	(a)	29.	(b)	49.	(d)	69	(b)	89.	(a)	109.	(c)	129.	(b)
10.	(b)	30.	(d)	50.	(c)	70.	(d)	90	(a)	110.	(c)	130.	(a)
11.	(a)	31.	(d)	51.	(a)	71.	(c)	91.	(a)	111.	(c)	131.	(b)
12.	(a)	32.	(c)	52.	(c)	72.	(b)	92.	(d)	112.	(a)	132.	(b)
13.	(c)	33.	(a)	53.	(d)	73.	(d)	93.	(a)	113.	(c)	133.	(c)
14.	(b)	34.	(c)	54.	(c)	74.	(c)	94.	(b)	114.	(d)	134.	(c)
15.	(a)	35.	(a)	55.	(a)	75.	(a)	95.	(b)	115.	(a)	135.	(b)
16.	(a)	36.	(d)	56.	(b)	76.	(c)	96.	(b)	116.	(a)	136.	(b)
17.	(b)	37.	(b)	57.	(c)	77.	(b)	97.	(c)	117.	(c)	137.	(b)
18.	(d)	38.	(d)	58.	(d)	78.	(b)	98.	(c)	118.	(a)	138.	(c)
19.	(d)	39.	(b)	59.	(d)	79.	(d)	99.	(b)	119.	(a)	139.	(a)
20.	(b)	40.	(a)	60.	(a)	80.	(b)	100.	(c)	120.	(b)	140.	(d)

Chapter-6

Storage Engineering

ONE LINER:

- In shallow bin, plane of rupture meets the grain surface before it strike the opposite side.
- As the increase in moisture content, the angle of repose also increased.
- Hermetic sealed units are more suited for cold storage.
- During storage, lateral pressure increase constantly as the bin is being filled.
- Cooling stored grain to prevent mold growth and insect activity includes removal of natural heat and heat from artificial drying.
- Henderson equation is used to determine the EMC of grains in percent dry basis.
- Lateral pressure along a bin wall is zero at the top and maximum at the bottom.
- Safe storage MC is always not equal to EMC of that product.
- Proper O_2 and CO_2 levels for the product and storage temperature may double the normal refrigerated storage life of the product.
- Seed grain between 7-9% moisture content should be packed in moisture resistant containers.
- Waxing and polishing enhance the appearance but do not improve the quality of the product.

- ❒ Deterioration of oil seed occur in storage and loss of viability is of little consequence in seed to be processed.
- ❒ The lateral pressure in grain storage is calculated by using hydrostatic formula.
- ❒ In granular material, internal friction is mainly due to cohesion.
- ❒ Seed is protected from increase in moisture content under low relative humidity conditions.
- ❒ Form Fill and Seal' machine is employed for filling liquid food products in flexible pouches.
- ❒ The natural air drying which may require several weeks in drying, are useful particularly to dry grains for short term storage in the farm.
- ❒ In deep bins, a part of the load is resisted by friction between the material stored and the side walls of the silo.
- ❒ One tonne of refrigeration is equal to 3000 kcal/hour.
- ❒ Package should be easily disposable, compatible with product and not be costly.
- ❒ Spherical-shape of glass containers is best to resist internal pressure.
- ❒ During storage of grains, insect development is enhanced by high moisture content.
- ❒ Bags filled with the grain shall be stacked 10 m away from the walls.
- ❒ The distance between two stacks should be 2m.
- ❒ High temperature and moisture leads to higher respiration of the grain.
- ❒ Aeration is carried out to remove moisture gradient in stored grains.
- ❒ Main reasons to use packaging are protection of the products from biotic and abiotic factors and also to enhance the shelf life of the product at normal temperature.
- ❒ Potatoes stored at 4°C and 90% RH are good for table purpose.
- ❒ Aeration of high moisture grains should not be carried out if the RH of ambient air is more than 70 %.
- ❒ In MA storage, the gaseous composition around the produce changed.
- ❒ Vegetables should be pre-cooled immediately after harvesting to remove field heat.
- ❒ Conditioning of the grains after drying is must for uniform moisture distribution.

- Storage tanks used in dairy plants maintain the milk at a temperature of 7°C or below.
- The microbial activity decrease with decreasing temperature below 0°C.
- In grain storage, deterioration index of zero corresponds to relative humidity of 30%.
- Handling and storage losses are less in bulk storage than bag storage of grains.
- Airy's theory is used to calculate lateral pressure for shallow bins.
- Storage above 16°C drastically increase respiration rate with consequent loss in dry matter.
- Proper packaging is essential for preserving freshness and nutrition of fruits and vegetables.
- A high R.H. of about 90% is desirable for most vegetables to prevent moisture loss and wilting.
- For the selection of strapping materials, the properties to be considered are tensile and elongation.
- MRTP act stands for Monopolies-Restrictive trade Practices Act; FPO stands for fruit product order.
- HMHDPE stands for High molecular High density Polyethylene; HIPS stand for High impact polystyrene; EVA stands for ethylene vinyl acetate copolymer
- Aseptic packaging is an essential part of UHT processed milk to give a commercially sterile product.
- Packaging material suitable for packaging of moisture sensitive food products must have low water vapor transmission permeability.
- A multilayered packaging material is used in packaging of UHT milk. Two materials used in making this composite material are polyethylene and aluminum foil.

Questions

1. **The average percentage loss of world food production due to insects during storage is about**
   (a) 1-5% (b) 5-10%
   (c) 10-15% (d) >20%
2. **An average one rat consumes ....kg of grain in year**
   (a) 10 (b) 25
   (c) 50 (d) 100
3. **In construction of rat proof godown, the plinth is constructed about .... m above the ground level**
   (a) 1 (b) 2
   (c) 2.5 (d) 2.5-3.5
4. **The shape of plinth in rat proof godown is**
   (a) S Shape (b) R Shape
   (c) L Shape (d) X Shape
5. **The chemical used to kill the rats and mice is called**
   (a) Fungicide (b) Rodenticide
   (c) Both (a) & (b) (d) None of the above
6. **.................... is a type of traditional storage structure**
   (a) Morai (b) Bukhari
   (c) Kothar (d) All of the above
7. **Morai type storage structure are used at .... region of India**
   (a) Southern (b) Northern
   (c) Eastern (d) Both (a) & (c)
8. **The capacity of morai storage structure is .... tonnes**
   (a) 9-35 (b) 3.5-18
   (c) 35-48 (d) 48-60

9. **The shape of morai storage structure is ....**
   (a) Cone (b) Sphere
   (c) Inverted truncated (d) Circular

10. **The capacity of bukhari type storage structure is .... tonnes**
    (a) 2-3 (b) 3.5-18
    (c) 18-21 (d) 22.5-42.5

11. **Kothar type storage structure are used at .... regions of India**
    (a) North (b) South
    (c) East (d) West

12. **The capacity of kothar storage structure is about .... tonnes**
    (a) 3.5-18 (b) 9-35
    (c) 35-50 (d) 50-75

13. **The capacity of mud kothi type storage structure is about .... tonnes**
    (a) 3.5-18 (b) 9-35
    (c) 50-120 (d) 1-50

14. **In which state of India the muda type storage structure are commonly used**
    (a) Telangana (b) Bihar
    (c) Karnataka (d) Tamilnadu

15. **The storage capacity of muda type grain storage structure is about .... tonnes**
    (a) 1-3 (b) 9-35
    (c) 3-18 (d) 35-50

16. **The shape of muda type storage structure is ....**
    (a) Sphere (b) Triangular
    (c) Cylindrical (d) All of the above

17. **Kanaj type storage structure are generally used in rural areas of ....**
    (a) Bihar (b) Karnataka
    (c) Delhi (d) Haryana

18. **Capacity of kanaj type storage structure is about .... tonnes**
    (a) 1-20 (b) 38-46
    (c) 3-5 (d) 20-28

19. **Kanaj type storage structure is made by**
(a) Iron rods (b) Bamboo splits
(c) Stainless steel (d) None of the above

20. **The shape of kanaj type storage structure is**
(a) Cylindrical (b) Conical
(c) Sphere (d) None of the above

21. **The walls of pusa bin is made of**
(a) Kachcha brick (b) Steel
(c) Stones (d) None of the above

22. **............plastic is used in pusa bin**
(a) 100 gauge (b) 350 gauge
(c) 500 gauge (d) 700 gauge

23. **The full form of CAP storage is**
(a) Cap and pole storage (b) Cover and plinth storage
(c) Cover and pole storage (d) Cap and plinth storage

24. **Which of the following type of cover used in CAP storage**
(a) 100 gauge (b) 500 gauge
(c) 300 gauge (d) 1000 gauge

25. **The storage capacity of CAP storage structure about .... tonnes**
(a) 1-50 (b) 150
(c) 250 (d) 1000

26. **The storage capacity of brick and cement bin is about .... tonnes**
(a) 1-10 (b) 50-80
(c) 1.5-60 (d) 100-150

27. **Moisture migration in stored grain mainly due to ....**
(a) Temperature change (b) Amount of stored grain
(c) Type of storage structure (d) None of the above

28. **The spoilage of grains takes at .... of the bin, in winter season.**
(a) Center (b) Top
(c) Bottom (d) Inclined

29. **In summer season the spoilage of grains takes at .... of the bin**
(a) Center (b) Top
(c) All of the above (d) Bottom

30. The plane of rupture meets the grain surface before it strikes the opposite side is known as ....

(a) Shallow bin (b) Both (a) & (c)

(c) Deep bin (d) None of the above

31. The plane of rupture meets the opposite side before its emerges from the grain is called ....

(a) Shallow bin (b) Deep bin

(c) None of the above (d) Both (a) & (b)

32. Hydraulic radius (R) of circular shape bin =

(a) d/4 (b) d/16

(c) d/2 (d) d/8

33. .... represents deep bin

(a) H<4R (b) H>4R

(c) H>=4R (d) H=4R

34. Which of the following represents shallow bin

(a) H>4R (b) H<4R

(c) H>4R (d) H=4R

35. Rankine's formula used for determination of pressure in .... bins

(a) Shallow bin (b) Both (a) & (b)

(c) Deep bin (d) None of the above

36. Janssen's formula used for determination of pressure in .... bins

(a) Shallow bin (b) Deep bin

(c) Both (a) & (b) (d) None of the above

37. The air flow rate for cooling of dry grains by aeration in silo is .... litre/ second/tone

(a) 0.8 (b) 1.6

(c) 3.2 (d) 4.8

38. The maximum dust velocity during aeration of grains in silo is ....m/s

(a) 3 (b) 6

(c) 10 (d) 15

39. ....fumigants used in sealed vertical storages

(a) Carbon dioxide (b) Methyl bromide

(c) Ethyl chloride (d) None of the above

40. Carbon dioxide generally kills insects .... then nitrogen

(a) Faster (b) Slower

(c) Equlal (d) None of the above

41. Generally the insects are killed when oxygen in the inter granular air falls about ....

(a) 4% (w/v) (b) 2% (w/w)

(c) 2% (w/v) (d) 4% (w/w)

42. In which of the following case the complete mortality occurs

(a) 20-24% R.H; 6.4%O_2 & 4.3 % CO_2

(b) 20-24% R.H; 3.2%O_2 &4.3 % CO_2

(c) 40-48% R.H; 6.4%O_2 &4.3 % CO_2

(d) 40-48% R.H; 3.2%O_2 &10.5 % CO_2

43. In godown the .... amount of extra space provided for inspection etc.

(a) 10% (b) 45%

(c) 30% (d) 60%

44. Control atmosphere storage conditions for apple is

(a) 1.5-10% CO_2, & 2.5 % O_2 (b) 1.5-10% CO_2, & 5 % O_2

(c) 10-15% CO_2, & 2.5 % O_2 (d) 10-15% CO_2, & 5% O_2

45. Which of the following conditions prevents the potato from sprouting

(a) 10-15 °C & 85-90 % R.H (b) 10-15 °C & 65-70 % R.H

(c) 4-6 °C & 65-70 % R.H (d) 4-6 °C & 85-90 % R.H

46. Recommended perforation of aeration in grain storage shall be ....

(a) 8% (b) 25%

(c) 15% (d) 5%

47. Perforation in duct for aeration should be uniformly spaced with minimum area about ....

(a) 10% (b) 5%

(c) 15% (d) 20%

48. Respiration rate is .... for old grain and fest infested grain as compared to fresh grain

(a) High (b) Equal

(c) Low (d) None of the above

49. The respiration rate of stored grain is mostly depends on

(a) Temperatures (b) Both (a) and (c)

(c) Moisture content (d) None of the above

50. The moisture content at which the respiratory rate undergoes a sharp change is called

(a) Crucial moisture content (b) Equilibrium moisture content

(c) Free moisture content (d) All the above

51. The capacity of khani (or) Patra storage structure is about .... tonne

(a) 5-10 (b) 2-3

(c) 10-155 (d) 15-30

52. Khani type storage structures used is ....States in India

(a) Rajasthan (b) Karnataka

(c) U.P (d) Orrissa

53. Patra type storage structures used in .... states in India

(a) Rajasthan (b) A.P

(c) U.P (d) Orrissa

54. Which of the most quite grain storage structure in India.

(a) Mild steel bin and RCC bin

(b) Plain or corrugated galvanized sheet bin

(c) Fiber glass and asbestos sheet bin

(d) Aluminum sheet and mild black sheet

55. A good quality storage structure should be

(a) Gas sealable for fumigation (b) Structurally adequate

(c) Durable and hygienic (d) All of the above

56. The storage loss of grains stored in bunker storage structure can be reduced up to .... by controlling insects and moisture level.

(a) 10-15% (b) 5- 10%

(c) 0-0.5% (d) 1-3%

57. The granular material behave as composite mass having the characteristic of

(a) Solid and gas (b) Liquid and gel

(c) Semi solid and semi liquid (d) Liquid and solid

58. At what temperature, potato can be stored for longer time ....

(a) 0 °C (b) 4 °C

(c) 10 °C (d) 16 °C

59. The recommended temperature for grape storage is

(a) 20 to 25°C (b) 10 to 15°C

(c) 5 to 10°C (d) -1 to 10°C

60. For safe storage of wheat maximum moisture content (wet basis) is ....

(a) 22% (b) 13%

(c) 28% (d) 18%

61. In EMC, during the seed storage, vapor pressure of grain is

(a) > atm (b) < atm

(c) 5 times > atm (d) =atm

62. In grain storage structures, rat proofing cones are provided at a height of

(a) 0.9 m (b) 1.2 m

(c) 1.5 (d) 1.8 m

63. For smooth floor surfaces, in vertical silo the slope angle should be

(a) 30-35° (b) 20-25°

(c) 35-40° (d) 25 -30°

64. The best method of estimating the pressure loss in fluid flow through granular solids is to use

(a) Blake-Kozney equation (b) Ergon equation

(c) Fourier equation (d) Henderson equation

65. Unit of volume is

(a) Pascal (b) Kg

(c) Bushel (d) None of these

66. The grain temperature has to be increased to more than .... to kill the insects

(a) 60 °C (b) 45 °C

(c) 35 °C (d) None of above

67. In summer/ winter the storage temperature effect the grains up to .... cm

(a) 100 cm (b) 200 cm

(c) 300 cm (d) None of these

68. Hecto-litre weight is a measure of

(a) Grain bulk density
(b) Grain true density
(c) Grain specific gravity
(d) None of the above

69. Fungi do not at ………. storage temperature.

(a) <0 °C
(b)-10 °C
(c) <20 °C
(d) >20 °C

70. A minimum thickness of silo wall kept as

(a) 5 cm
(b) 15 cm
(c) 10 cm
(d) 20 cm

71 Safe storage temperature for apple in cold storage is

(a) 2-3 °C
(b) 3-4°C
(c)-2-(-1)°C
(d) -3-(-2)°C

72. The chilling injury of banana occurs at a temperature below

(a) 4-5 °C
(b) 8-10 °C
(c) 12-13 °C
(d) 14-15 °C

73. 1TR is equal to

(a) 3024 kcal/h
(b) 50.5 kcal/s
(c) 210 kcal/min
(d) All the above

74. The usual temperature maintained in storage of fruits and vegetables lies within …°C

(a)-25 to -29 °C
(b) 0- 10 °C
(c) -16 to -18 °C
(d) 10-20 °C

75. Concentration of CO_2, and O_2, in CA storage for apple should be maintained at

(a) 1.5 to 10% & 2.5%
(b) 5 to 10% & 3%
(c) 5 to 10% & 5%
(d) 5 to 6% & 10%

76. …………………is not a solid fumigant

(a) Celphos
(b) Phostoxin
(c) Methyl bromide
(d) Delta

77. Controlled atmosphere storage is use for storage of……………..

(a) Juices
(b) Jams
(c) Dried products
(d) Fresh fruits and vegetables

78. The ratio of height to dimeter of squat soli is ....

(a) >0.8 (b)<=0.5

(c) >1.2 (d) = 1.2

79. Flow factor in a storage bin is a function of ....

(a) Type of material (b) Hopper of bin wall

(c) Both (a) & (b) (d) All the above

80. .... is a method of fruits and vegetable processing

(a) Dry cleaning (b) Wet cleaning

(c) Both (a) & (b) (d) None of these

81. Grain is spoiled in bin due to accumulation of

(a) Temperature (b) Moisture

(c) Heat (d) Relative humidity

82. Shelf life of fruits and vegetables stored in CAP storage can be increased by ......

(a) Increase CO_2, level (b) Decreasing the temperature

(c) Decrease O_2, level (d) All of these

83. The generation of heat in stored food grains is due to

(a) Sensible heat (b) Heat of respiration

(c) Specific heat (d) Both (a) and (b)

84. The heat load in cold storage system are classified like

(a) Freezing load (b) Chilling load

(c) Cooling load (d) All of the above

85. The function of air tight storages is to

(a) Protect the wet grain from mold attack

(b) Arrest the insect infestation

(c) Decrease the oxygen content

(d) Both (a) and (b)

86. Methyl bromide and phosphine are used as

(a) Fungicide (b) Fumigant

(c) Pesticide (d) Germicide

87. Morai structure is generally suitable for

(a) Mustard, soybean and wheat (b) Maize, bazar and oat

(c) Paddy, potato and tobacco (d) Paddy, maize and sorghum

88. ....is not a liquid fumigant

(a) Phostoxin (b) Ethylene dibromide

(c) Ethylene dichloride (d) Carbon tetrachloride

89. .... represents the aseptic processing and packaging

(a) Sterilized food stored in raw containers

(b) Raw food stored in raw containers

(c) Sterilized food stored in pre sterile containers

(d) None of these

90. ....chemical is used to prevent the sprouting of potato

(a) Ethylene (b) O_2

(c) NH_3 (d) CO_2

91. The concentration of chlorinated water is used for washing of fruits and vegetables is

(a) 100 to 150 ppm (b) 150 to 300 ppm

(c) 10 to 20 ppm (d) 400 to 600 ppm

92. Which of the following method of preservation keeps the fruits and vegetables in fresh like form?

(a) De-hydro freezing (b) Micro wave heating

(c) Freezing (d) Freeze drying

93. Wax coating is done on fruits and vegetables to retard

(a) Retard respiration (b) Enhance appearance

(c) Retard dehydration (d) All of the above

94. Which of the following packaging film can effectively use for MAP

(a) Polyproplene (b) HDPE

(c) PVC (d) All the above

95. Which is one of the best packing materials to inhabit moisture migration from the packaged products

(a) HDPE (b) Laminated foil

(c) Steel (d) None of these

96. As per definition given by Issacson and boyd, deep bins is calculated by

(a)$Hd/D=0.75\ (K/\mu)$ (b) $Hd/D<0.75\ (1/\mu K)$

(c) $Hd/D-0.75\ (1/\mu K)$ (d) $Hd/D=0.75\ (\mu/K)$

97. ....is the effective method for cool the grains in bins and slow down insect activity

(a) Aeration
(b) Fumigation
(c) Both (a) & (b)
(d) None of these

98. The most common air flow rates for aerating paddy range varies from

(a) 0.07-0.28 m^3/min/tonne
(b) 0.80-1.00 litre/sec/tonne
(c) 0.01-0.04 m^3/min/tonne
(d) 1.20-1.60 litre/sec/tonne

99. The ratio of lateral pressure to vertical pressure for calculating later pressure in deep bin using Janssen's equation is

(a) Constant
(b) Decrease with loading and unloading
(c) Increase with loading and unloading
(d) All the above

100. To avoid spoiling, silage should be removed at a rate of

(a) 20 cm/day
(b) 5 cm/day
(c) 15 cm/day
(d) 10 cm/day

101. Pit silos have

(a) Trapezoidal cross section
(b) All of these
(c) Rectangular cross section
(d) Circular cross section

102. Most fruits and vegetables are stored at

(a) Low temperature and low humidity
(b) High temperature and low humidity
(c) Low temperature and high humidity
(d) High temperature and high humidity

103. Permeability of gases and vapour through a packaging film depends on;

(a) Concentration difference between two sides of the film
(b) Thickness of the film
(c) Thermal conductivity of film
(d) Area of film surface

104. Hermetically sealed containers are essential for

(a) Vacuum and pressure packaging
(b) Aseptic packaging
(c) Flexible packaging
(d) Controlled atmosphere packaging

105. Modified humidity packaging (MHP) systems are designed to control

(a) CO_2 and N_2 (b) Dehydration and condensation

(c) Oxygen (d) All the above

106. Polyethylene films are good barrier of:

(a) Water vapour (b) Aroma

(c) Oxygen (d) All the above

107. .... is used both as contact chemical as well as fumigant

(a) Phostoxin (b) Detia

(c) Dichlovros (c) Ethylene dichloride

108. Short term storage of seed under dry conditions retains good quality in

(a) Tin containers (b) Plastic containers

(c) Porous paper (d) Metal containers

109. Garlic bulbs before storing then in an ordinary room are used in shady places for

(a) 3-4 days (b) 7 days

(c) 4-8 days (d) 15 days

110. Optimum temperature and relative humidity for tomato storage

(a) 8 °C and 90 to 95% (b) 2 °C and 90%

(c) 8 °C and 95 to 100% (d) 2 °C and 100%

111. ....can be stored at low temperature and low RH?

(a) Carrots (b) Potatoes

(c) Onions (d) All the above

112. Which of the following cannot be stored safely in high humidity environment?

(a) Tomato (b) Potatoes

(c) Apple (d) Onion and garlic

113. Safe storage recommended temperature of grapes is

(a)-1 to 10 °C (b) 10 to 15 °C

(c)-5 to 0 °C (d) 15 to 20 °C

114. The usual temperature maintained in frozen storage of fruits and vegetables lies within...........°C

(a)-25 to-29°C (b)-16 to-18°C

(c) 4 to 8°C (d) 8 to 20°C

115. The chilling of fruits and vegetables involve bringing down the temperature of commodities to

(a)-4 to 4°C (b)-1 to 8°C

(c) 4 to 12°C (d) 10 to 18°C

116. Safe storage temperature for apple in cold store is

(a) 2 to 3°C (b) 3 to 4°C

(c) -2 to-1°C (d)-3 to-2°C

117. The latent heat of fusion for water is

(a) 540 kJ/kg (b) 80 kJ/kg

(c) 540 kcal/kg (d) 335 kJ/kg

118. The latent heat of evaporation for water is

(a) 540 kJ/kg (b) 80 kJ/g

(c) 335 kJ/kg (d) 540 kCal/kg

119. The normal heat generation by respiration for grains is usually

(a) 0.01 MJ/tonne-day for dry grain

(b) 0.05 MJ/tonne-day for dry grain

(c) 0.25 MJ/tonne-day for dry grain

(d) 1.05 MJ/tonne-day for dry grain

120. The heat generation by respiration for fruits and vegetables at 0°C is usually

(a) 0.005 to 0.05 MJ/tonne-day (b) 0.05 to 0.5 MJ/tonne-day

(c) 0.5 to 10 MJ/tonne-day (d) 10 to 100 MJ/tonne-day

121. The fruits in which there is a short but abrupt increase in the rate of respiration, which occurs near the point of optimum ripeness, is known as

(a) Temperate fruit (b) Climacteric fruit

(c) Non-climacteric fruit (d) None of the above

122. .... is a climacteric fruit?

(a) Grape (b) Avocado

(c) Lemon (d) Pineapple

123. .... is a symptom of chilling injury?

(a) Internal and external browning (b) failure to ripen

(c) blemished skin (d) all of these

124. The chilling injury of banana occurs at a temperature below

(a) 8-10°C (b) 14-15°C

(c) 4-5°C (d) 12-13°C

125. Freezer burn is a storage defect in food products which is caused in

(a) dehydration (b) concentration

(c) frozen storage (d) refrigeration

126. Excessive shrinkage and early sprouting occur if potatoes are held at temperatures above ....°C

(a) 3 (b) 10

(c) 20 (d) 40

127. The optimum storage condition for potato are

(a) 3-10°C and 90-98% RH (b) 10-20°C and 90-98% RH

(c) 3-10°C and 80-85% RH (d) 10-20°C and 80-85% RH

128. The process of exposing the food to either electromagnetic or ionizing radiations to destroy the microorganisms is known as

(a) irradiation (b) dehydration

(c) sterilization (d) pasteurization

129. Altering environmental conditions to prevent growth of nilcro-organisms can help preserve food. Such conditions are called .... conditions.

(a) bacteriodyanamic (b) bacteriostatic

(c) bacteriocidal (d) None of the above

130. Methods in which most of the microorganisms present in the food are killed as in canning, cooking, irradiation etc. are ....conditions.

(a) bacteriodyanamic (b) bacteriostatic

(c) None of the above (d) bacteriocidal

131. .... involves preservation with help of microorganisnms?

(a) aseptic processing (b) pasteurization

(c) fermentation (d) None of the above

132. Which of the following preservation methods maintains the texture and original characteristics of fruits and vegetables to the closest possible level?

(a) chilling (b) drying

(c) fermentation (d) pasteurization

133. .... is a heat processing method of food preservation?

(a) carbonation (b) flash pasteurization

(c) evaporation (d) None of the above

134. The normal level of sodium benzoate for use in food preservation is

(a) 0.01-0.03% (b) 0.03-0.2%

(c) 3-5% (d) 0.5-3%

135. Sheet test is used for measurment of

(a) Gel consistency (b) Porosity

(c) Viscosity (d) None of the above

136. Lemon, mango and other such pickles owe their keeping quality partly to the large amount of .... added in the preparation.

(a) Sugar (b) Chemicals

(c) Acid (d) Salt

137. The squash has a minimum TSS value of .... %.

(a) 65 (b) 40

(c) 30 (d) 10

138. During bottle filling, an empty space of ...... is kept on the fruit squash.

(a) 2.5-3 cm (b) 1-2 cm

(c) 4-5 cm (d) 0.1-0.5 cm

139. The RTS fruit drink Should have .... % fruit juice and............ % TSS.

(a) 10, 20 (b) 10, 10

(c) 20, 20 (d) 20, 10

140. The usual acid content in RTS is

(a) 3% (b) 0.03%

(c) 0.3% (d) 5-6%

141. The nectar should have ....% fruit juice and .... % TSS.

(a) 10, 25 (b) 10, 15

(c) 20, 25 (d) 20, 15

142. The sherbet should have at least .... % TSS.

(a) 25 (b) 35

(c) 55 (d) 65

143. The product obtained after boiling of pulp with sufficient sugar to thick enough consistency and firm enough to hold tissues in position is known as

(a) Jam (b) Jelly

(c) Preserve (d) Marmalade

144. The same process of jam making is used for jellies except that .... is used in place of fruit pulps.

(a) Pectin (b) fruit juice

(c) Fruit rinds (d) chemicals

145. A mature fruit or vegetable, or its pieces, is impregnated with heavy sugar syrup until tender and transparent, is called

(a) Jam (b) marmalade

(c) Jelly (d) preserve

146. .... is important for preparation of pickles?

(a) Acetic acid (b) lactic acid forming bacteria

(c) Citric acid (d) None of the above

147. The lactic acid bacteria is most active at

(a) 10°C (b) 20°C

(c) 30°C (d) 40°C

148. Vinegar is also called

(a) Citric acid (b) Forming acid

(c) Acetic acid (d) None of the above

149. During preparation of vinegar, the sliced fruit and Sugar are kept in the ratio of:

(a) 1:1 (b) 3:1

(c) 2:1 (d) 4:1

150. If a product is preserved in vinegar, the final concentration of vinegar in terms of acetic acid should not be less than

(a) 1% (b) 2%

(c) 3% (d) 8%

151. On an average 10 kg tomato would yield..........kg juice.

(a) 5 kg (b) 9 kg

(c) 7 kg (d) 9.5 kg

152. For pulping, toinato is usually heated at .... for ....min.

(a) 40-50°C, 3-5 min (b) 70-90°C, 3-5 min

(c) 40-50°C, 30-50 min (d) 70-90°C, 30-50 min

153. The tomato ketchup should have more than ....% tomato solids and .... % total solids.

(a) 12, 25 (b) 25, 45

(c) 4, 12 (d) 25, 12

154. The tomato paste has minimum .... % solids.

(a) 15 (b) 33

(c) 25 (d) None of above

155. During canning of vegetables, the vegetables are heat treated for

(a) almost complete killing of micro-organisms

(b) killing of spoilage micro-organisms

(c) either (a) or (b)

(d) inactivation of enzymes.

Answer Key

1.	(b)	31.	(b)	61.	(d)	91.	(a)	121.	(b)	151.	(c)
2.	(a)	32.	(a)	62.	(b)	92.	(c)	122.	(b)	152.	(b)
3.	(a)	33.	(c)	63.	(a)	93.	(d)	123.	(d)	153.	(a)
4.	(c)	34.	(b)	64.	(b)	94.	(d)	124.	(d)	154.	(b)
5.	(b)	35.	(a)	65.	(c)	95.	(b)	125.	(c)	155.	(c)
6.	(d)	36.	(b)	66.	(a)	96.	(c)	126.	(b)		
7.	(d)	37.	(a)	67.	(c)	97.	(a)	127.	(a)		
8.	(b)	38.	(c)	68.	(a)	98.	(a)	128.	(a)		
9.	(c)	39.	(b)	69.	(a)	99.	(a)	129.	(b)		
10.	(b)	40.	(a)	70.	(b)	100.	(d)	130.	(d)		
11.	(a)	41.	(c)	71.	(c)	101.	(d)	131.	(c)		
12.	(b)	42.	(b)	72.	(c)	102.	(c)	132.	(a)		
13.	(d)	43.	(c)	73.	(d)	103.	(b)	133.	(b)		
14.	(b)	44.	(a)	74.	(c)	104.	(a)	134.	(b)		
15.	(a)	45.	(d)	75.	(a)	105.	(b)	135.	(a)		
16.	(c)	46.	(c)	76.	(c)	106.	(a)	136.	(d)		
17.	(b)	47.	(a)	77.	(d)	107.	(c)	137.	(b)		
18.	(a)	48.	(a)	78.	(b)	108.	(c)	138.	(a)		
19.	(b)	49.	(b)	79.	(c)	109.	(a)	139.	(b)		
20.	(c)	50.	(a)	80.	(c)	110.	(a)	140.	(c)		
21.	(a)	51.	(b)	81.	(b)	111.	(c)	141.	(d)		
22.	(d)	52.	(d)	82.	(d)	112.	(d)	142.	(d)		
23.	(b)	53.	(b)	83	(d)	113.	(a)	143.	(a)		
24.	(d)	54.	(a)	84.	(d)	114.	(b)	144.	(b)		
25.	(b)	55.	(d)	85.	(d)	115.	(b)	145.	(d)		
26.	(c)	56.	(c)	86.	(b)	116.	(c)	146.	(b)		
27.	(a)	57.	(c)	87.	(d)	117.	(d)	147.	(c)		
28.	(b)	58.	(b)	88.	(a)	118.	(d)	148.	(c)		
29.	(d)	59.	(a)	89.	(c)	119.	(a)	149.	(b)		
30.	(a)	60.	(b)	90.	(a)	120.	(c)	150.	(b)		

Chapter-7

Processing Technology

ONE LINER:

- ❐ In decimal reduction time, the total microbial destruction is ensured.
- ❐ UHT processing is also known as aseptic processing.
- ❐ TDT curve is used in milk homogenization.
- ❐ Blanching is generally not followed in vegetables.
- ❐ TDT curve is drawn between temperature vís log (survivors).
- ❐ Pasteurization kills all microorganisms.
- ❐ The batch pasteurizers are costly.
- ❐ HTST pasteurization system adapts itself well to CIP cleaning.
- ❐ Pasteurization of milk destroys all micro-organisms.
- ❐ The thermal death time increases with an increase in process temperature.
- ❐ Wet steam contains no water droplets or mist.
- ❐ Canning involves placing a product in a container, sealing it and heating- the sealed container under pressure.
- ❐ During pasteurization, temperature is maintained below 100°C.
- ❐ No loss of water soluble materials takes place during immersion blanching process.
- ❐ The prime objective of sterilization is to kill pathogenic microorganisms.

- The main purpose of pasteurization is to destroy pathogenic organism.
- D-value is independent of temperature and type of micro-organism.
- Hydrostatic sterilizer is used for sterilization of canned food.
- UHT is used in aseptic processing.
- Microbial destruction is a first order reaction.
- Pressure vessels or retorts are used for canning.
- Spore-forming bacteria are more heat resistant than vegetative bacteria.
- Pasteurization does not kill all micro-organisms.
- The HTST pasteurization system adopts itself well to CIP system.
- The shelf life of food material is increased by increasing microorganisms, chemical reactions and biochemical reactions.
- D-value (decimal reduction time) depends upon temperature.
- Bottle sterilization of milk is at 105-110°C for 30-45 minutes.

Questions

1. **Pasteurization destroys .... microorganism**

   (a) Only pathogenic microorganisms

   (b) Only spoilage microorganisms

   (c) Only heat resistant microorganisms

   (d)All types of microorganisms

2. **Presumptive test is found positive for:**

   (a) B.Col (b) E. Coli

   (c) Lactofacillus Bulgaricus (d) Clostridium Botulinum

3. **99.999% reduction in microbial population (1 survivor in 100,000) is equivalent to:**

   (a) 4D (b) 5D

   (c) 100,00OD (d) 0.00001D

4. **The relation between D-value in thermal death kinetics and lethality (S) is:**

(a) S = F/D (b) S = FD

(c) S = F- D (d) D = FS

5. **Microorganisms are destructed by heat at fixed temperature follows:**

(a) Third order reaction (b) Second order reaction

(c) First order reaction (d) Zero order reaction

6. **The cooled milk will leave at ....temperature during Regeneration section has 95% regeneration efficiency.**

(a) 13.1°C (b) 15°C

(c) 10.8°C (d) 12°C

7. In thermal processing of food destruction of microorganisms is governed by an equation having the following order:

(a) Zero (b)First

(c) Second (d) Third

8. **The decimal reduction time for an organism at 250°F is denoted by:**

(a) Z (b) F

(c) F_0 (d) None of the above

9. **Generally, suitable process for heat sterilization is**

(a) Low acid food (b) High acid food

(c) High proteins foods (d) None of the above

10. **At the time of fruit juice canning pasteurization is done at the temperature**

(a) 80°C (b) 77°C

(c) 74°C (d) 71°Cc

11. **The D-value is the time required to destroy................ % of the microorganisms:**

(a) 10 (b) 89

(c) 90 (d) 100

12. **D-value signifies, when thermal destruction of micro-organisms**

(a) Doubling time (b)Decimal reduction time

(c) Regeneration time (d) None of the above

13. D-value (decimal reduction time) depends upon:

(a)Temperature

(b) Heating time

(c) Number of microorganisms initially present in the product

(d) Death rate constant

14. If milk enters the regeneration section at 10°C having 95% regeneration efficiency, it will leave the regeneration section at (assuming 72°C as pasteurization temperature):

(a) 72°C (b) 71°C

(c) 68.9°C (d) 9.5°C

15. Z-value of the processor are commonly expressed in:

(a) °C or °F (b) Percent

(c) Second or minute (d) Number of micro-organism

16. A process that reduces the micro-organisms by 8 decimal reduction applied to raw material which contains 10^5 spores would reduce microbial number to:

(a) 10^{-2} (b) 10^{-3}

(c) 10^{-5} (d) 10^{-1}

17. In thermal processing, the destruction of all micro-organisms in food by

(a) Sterilization (b) Scalding

(c) Blanching (d) Pasteurization

18. Microbial destruction time is inversely proportional to decimal reduction time,

(a) Z value (b) F-value

(c) F value (d) none

19. The rate of reduction is a first order reaction that is when food is heated to a temperature is high enough to destroy contaminating micro-organisms, the same percentage die in a given time interval regardless of the numbers present initially. This is known as

(a) Z-value (b) F-value

(c) D-value (d) Logarithmic order of death

20. Pasteurization of fruit juice-canning is done at the temperature

(a) 81°C (b) 77°C

(c) 74°C (d) 71°C

21. Use of subatomic particies in which beta or gamma rays are used for treating the product so that ionization of the microorganism occurs is called:

(a) Cold pasteurization (b) Cold sterilization
(c) Both (a) & (b) (d) None

22. In order to make canned food (pH > 4.5) safe, heat processing must reduce the concentration of the spores of Clostridiun botulinum by:

(a) 15D (b) 12D
(c) 10D (d) 8D

23. The cans are cooled rapidly after processing at .... temperature

(a) 10°C (b) 39°C
(c) 49°C (d) 25°C

24. The bacteria can be destroyed by:

(a) Ultra-violet rays (b) Extreme heat
(c) Sun light (d) All the above

25. Compared to conventional sterilization, the reduction in enzymatic activity in UHT processing is:

(a) Less (b) More
(c) Equal (d) None of these

26. Microorganisms destroyed from food stuff during microwave heating is due to:

(a) Generation of heat
(b) Ionic polarization
(c) Susceptible to microwave radiation
(d) Dipole rotation

27. During sterilization of filled cans in which convection is the dominant mode of heat transfer, the slowest heating point in the container is usually taken as:

(a) Geometric centre of can (b) Top layer of liquid
(c) One third from bottom (d) Bottom layer of liquid

28. In milk processing, the lower temperature longer time process operating at 63°C for 30 minutes causes greater change to flavor and slightly greater loss of vitamins than HTST method where milk is processed at:

(a) 72°C for 15 seconds (b) 94°C for 3 second
(c) 100°C for I second (d) 90°C for 5 second

29. After canning, cooling the can will:

(a) Cook the contents (b) Stop the cooking process

(c) Stack burn (d) Discolour the contents

30. Typical temperatures for a range of food processing operations are indicated below.

(a) Spray drying=200-250 °C (b) Extrusion cooking = 140-180 °C

(c) Canning = 110-125 °C (d) Hot air drying = 50-100 °C

31. During thermal processing of a liquid food, if data is collected on number of survival organism (N) per ml of the product at various instants of time (1) and if a graph is plotted between log N and , the relationship obtained will be:

(a) Hyperbola (b) Parabola

(c) Exponential (d) Straight line

32. The time required to kill all mesophilic spores that can cause food spoilage is referred to as

(a) Residence time (b) Thermal death time

(c) Pasteurization time (d) None of the above

33. The temperature range during ultra-high-temperature (UHT) sterilization of milk is.............

(a) 90 to 100°C (b) 100 to 115°C

(c) 135 to 150°C (d) None of the above

34. Food material should be thermally processed based on its heat sensitivity,

(a) Vegetative cells of fungi (b) Spores of bacteria

(c) Spores of fungi (d) Vegetative cells of bacteria

35. Irradiation, pulsed electric field and high-pressure methods of food processing preservation can be categorized as

(a) Minimal processing (b) Thermal processing

(c) Non-thermal processing (d) Chemical processing

36. Generally, Fruits require lower thermal processing time than vegetables owing to fruits:

(a) Low pH and high moisture (b) Low pH and high sugar

(c) Low pH and low moisture (d) Low pH and low sugar

37. Commercial sterility is assured by...................testing in aseptic processing and packaging of the food

(a) Large and statistically random samples

(b) Small sample

(c) Small but statistically random samples

(d) Large samples

38. Which method is generally used on large scale to protect milk against bacteria infection

(a) Pasteurization (b) Evaporation

(c) Canning (d) Dehydration

39. The transformation of product at the time of extrusion is:

(a) Reversible

(b) Irreversible

(c) Reversible as well as irreversible

(d) None of the above

40. Regeneration is economical when the product is

(a) Heated (b) Cooled

(c)Heated and cooled (d) None of the above

41. Clostridium botulinum is a:

(a) Scientific name of a plant (b) Insect

(c) Microorganism (d) Chemical

42. There is only one difference between a pasteurizer and a sterilizer,

(a) Temperature attained (b) Heating agent used

(c) Design of equipment (d) None of the above

43. In thermal processing, the destruction of all micro-organisms in food is ....

(a) Pasteurization (b) Sterilization

(c) Blanching (d) Scalding

44. Losses of nutrients during heating of food products are guided by an equation of the following order:

(a) Zero (b) First

(c) Second (d) None of the above

45. If a can has a spore load of 100 having to Do value of the (b) specific spore as 1.6 min, the target process time so as to get a final spore load 10^{-5} (1 in 100000) will be

(a) 9.6 min (b) 11.2 min

(c) 12.8 min (d) None of the above

46 The process time at 121°C for 10 log cycle reduction of Clostridium botulinum (D_{121} =0.216 min) will be

(a) 2.16 min (b) 21.6 min

(c) 216 min (d) None of the above

47. In most ideal circumstances, a bacteria can multiply as many times as:

(a) Every 20 to 30 seconds (b) Every 20 to 30 minutes

(c) Every 20 to 30 hours (d) Every 2 to 3 days

48. Thermal vapor re-compression is used in

(a) Evaporator (b) Pasteurizer

(c) Homogenizer (d) None of the above

49. Sterilization of milk is carried out to the extent of destroying

(a) Only spoilage organisms (b) Only pathogenic organisms

(c) All types of organisms (d) Only heat resistant organisms

50. Before being allowed to pass through the pasteurizer, milk and fruit juice are de-aerated. This is done for the following reasons:

(a) Reduce fouling of pasteurizer (b) Reduce oxidative deterioration

(c) Increase rate of heat transfer (d) All the above

51. For any thermal process the decimal reduction time is inversely related to a function of the process:

(a) Temperature (b) Time

(c) Pressure (d) None of the above

52. Loss of nutrients during thermal processing of food is guided by an equation of the following order:

(a) Zero (b) First

(c) Second (d) None of the above

53. In microbial destruction, the decimal decrease time is inversely proportional to:

(a) Z value (b) Universal gas constant

(c) Initial concentration (d) Reaction rate

54. Blanching is the technique of heating food to a predetermined temperature in order to achieve:

(a) Destruction of pathogenic bacteria

(b) The extraction of moisture of the product

(c) Gelatinization of starch globules

(d) Inactivation of enzymes and destruction of peroxides

55. Inactivation of enzymes is referred to as:

(a) Leaching (b) Pasteurization

(c) Blanching (d) Sterilization

56. The thermal destruction of micro-organisms can be represented as (No = number of organisms initially present, = number or viable organisms at any instant of time. t and is a constant)

(a) dN/dt=-KN (b) dN/dt = KN

(c) First order reaction (d) None of these

57. The pH range that bacteria prefer to grow in is:

(a) Less than 4.5 (b) More than 8

(c) 6.7 to 7.5 (d) None of the above

58. Thermal death time (TDT) curve is plotted between:

(a) Time v/s temperature

(b) Temperature v/s log (time)

(c) Time v/s log (number of survivors)

(d) None of the above

59. Foods with low acidity, such as fish, poultry, pork, and most vegetables, must be prepared at high temperatures.

(a) Lower than 100°C (b) Higher than 100°C

(c) Higher than 88°C (d) Equal to100°C

60. Larger decimal reduction time implies:

(a) Micro-organisms more heat resistant

(b) Micro-organisms less heat resistant

(c) Microbial growth reduces with time

(d) No effect on micro-organisms

61. Pasteurization of milk is necessary for:

(a) Long shelf life at room temperature

(b) Destruction of all microorganisms

(c) Destruction of only pathogenic microorganisms

(d) Long shelf life under refrigeration

62. Which of the following is non-thermal processing technique

(a) PEF (b) HPP

(c) UV-C (d) All of these

63. In High Pressure processing technique, which fluid is used as pressure transmitting fluid

(a) Water (b) Oil

(c) Both (a) & (b) (d) None of the above

64. D value depends on

(a) Number of microorganisms present in food

(b) Time of heating

(c) Temperature of heating

(d) None of the above

65. The unit for radiation dose is

(a) KGy (b) Kg

(c) Pa (d) All Of these

66. The heavy dose irradiation is used in food

(a) <1KGy (b) 1-10KGy

(c) >10KGy (d) None of these

67. The medium dose irradiation is used in food

(a) <1KGy (b) 1-10KGy

(c) >10KGy (d) None of these

68. The low dose irradiation is used in food

(a) <1KGy (b) 1-10KGy

(c) >10KGy (d) None of these

69. An expression, which gives the rate of change in reaction or microbial spoilage with a change in 10°C temperature is

(a) z value (b) D value

(c) Q_{10} value (d) k value

70. The characteristics of microorganisms or vitamins as regards their stability during thermal processing are mentioned

(a) D value and z value
(b) D value and Q_{10} value
(c) z value and Q_{10} value
(d) Only D value

71. Appertisation is also known as

(a) Blanching
(b) Pasteurization
(c) Canning
(d) None of these

72. Before the actual processing activity, cans are passed through an exhaust box or steam chamber for 7-10 minutes during vegetable canning.

(a) Reduce the pressure
(b) Reduce the enzymatic activity.
(c) Remove the entrapped air from fruit tissues and the can.
(d) Remove the undesirable odour.

73. When a microorganism suspension is heated to a constant temperature, the reaction kinetics of the organism's number decreases as follows:

(a) Linear
(b) Hyperbolic
(c) Parabolic
(d) Exponential

74. Critical microorganism for canned foods is:

(a) Bacillhus stearothermophilus
(b) Yeasts and moulds
(c) Salmonella *spp.*
(d) Clostridium botulinum

75. Clostridium Botulinum is a common indicator bacterium in the following processes:

(a) Pasteurization
(b) Blanching
(c) Sterilization
(d) Drying

76. In HTST pasteurization, the milk is held at 72°C:

(a) 30 minutes
(b) 30 sec.
(c) 15 sec.
(d) 15 minutes

77. The D value represents the amount of time it takes to destroy the bacteria by:

(a) 10%
(b) 90%
(c) 100%
(d) 50%

78. Temperatures are frequently used in canning to kill spoilage organisms and inactivate enzyme activity.

(a) Below 100°C
(b) Above 100°C
(c) Below 80°C
(d) Above 80°C

79. Pressure vessels or retorts are used during

(a) Evaporation (b) Canning

(c) Dehydration (d) Extraction

80. The procedure for sealing the cans in an airtight manner is called as

(a) Seaming (b) Double seaming

(c) Lidding (d) Clinching

81. The practice of rapidly heating and cooling milk to control bacteria growth is known as:

(a) Blending (b) Regeneration

(c) Pasteurization (d) Evaporation

82. In laminar flow, the holding section of the HTST pasteurizer is the same length as in turbulent flow:

(a) More than (b) Less than

(c) The same (d) None of the above

83. UHT sterilization of milk occurs at temperature range of

(a) 90-100 °C (b) 100-113 °C

(c) 135-150 °C (d) 150-175 °C

84. Thermal process of conduction heating food material is based on the temperature at the

(a) Slowest heating point (b) Surface

(c) Point between surface and centre (d) None of the above

85. The temperature at which of the following is a temperature and time combination for sterilising vegetables

(a) 62°C for 15 min (b) 88°C for 15 min

(c) 100°C for 15 min (d) 121°C for 15 min

86. Very high dose of irradiation is used for

(a) Inhibition of germination, disinfection, ripening delay

(b) Shelf life extension, elimination of pathogens

(c) Reduce or eliminate virus contamination

(d) Sterilization

87. For sterilisation, rotary or reel sterilisers are utilised.

(a) Bulk materials (b) Canned food

(c) Liquid food in pipes (d) None of the above

88. Complete gelatinization during extrusion can be obtained at:

(a) 100°C and 24-27% m.c. (b) 100°C and 30-32% m.c.

(c) 110°C and 30-32% m.c. (d) 110°C and 24-27% m.c.

89. Which of the following sterilisation procedures has not been used for commercial sterilisation to achieve complete gelatinization during extrusion?

(a) Moist heat (b) Hydrogen peroxide

(c) Ionizing radiation (d) Dry heat

90. When applied to raw material containing 10 spores, a method that decreases microorganisms by 8 decimal reductions (8D) reduces the microbial number to

(a) 10^{-2} (b) 10^{-3}

(c) 10^{-5} (d) 10^{-1}

91. Booster pump is located just before:

(a) Regeneration section (b) Heating section

(c) Holding section (d) Cooling section

92. The z value concept is used in:

(a) Irradiation (b) Drying

(c) Sterilization (d) Freezing

93. The following are the temperature-time relationships for milk sterilisation in bottles:

(a) 140-150°C for 2-3 sec (b) 135°C for 30 sec

(c) 110-120°C for 10-20 minutes (d) 105-110°C for 30 to 45 minutes

94. During thermal processing, the change in the number of microorganisms (N) with time (0) is given by equation

(a) $\frac{N}{N_0} = e^{-k\theta}$ (b) $\frac{N}{N_0} = e^{k\theta}$

(c) $\frac{N}{N_0} = \text{In}(k\theta)$ (d) $\frac{N}{N_0} == \text{In}(-k\theta)$

95. The normal sequence followed during the operations of canning for fruits and vegetables is

(a) Lidding- double seaming- processing-cooling

(b) Double seaming- exhausting- heat processing- cooling- lidding

(c) Lidding- exhausting- double seaming- heat processing-cooling

(d) Exhausting- lidding- heat processing- double seaming- cooling

96. **Which of the following statement is false?**
   (a) The decimal reduction time is same for all gram D positive bacteria.
   (b) More decimal reduction time indicates higher thermal resistance of the bacteria.
   (c) The death rate curve, which is a plot of number of microorganisms-vs- processing time, is usually plotted on a semi log paper.
   (d) The decimal reduction time decreases with increase in processing temperature.

97. **Ohmic heating woks on the principle of**
   (a) Electrical resistance (b) Mechanical resistance
   (c) Electrical conductance (d) Thermal resistance

98. **Which of the following statement is not true?**
   (a) Washing of the vegetables can reduce the total processing time to reach a predefined level of commercial sterility.
   (b) If the raw product contains more number of micro organisms, the time of processing will be more to reach a certain predefined micro-organism level.
   (c) The commercial sterility is to bring the bacterial load to 1 in 108 cans.
   (d) The thermal death time curve is a plot of D values against temperature.

99. **The change in temperature required to bring about a 10 fold change in decimal reduction time is known as**
   (a) Thermal death rate (b) Z value
   (c) Q value (d) Q_{10} value

100. **If the D value of a microorganism is plotted with respect to time on a semi-log paper, then the slope of the line is**
   (a) Z (b) -z
   (c) 1/z (d) -1/z

101. **Microwave oven works on**
   (a) Dielectric (b) Thermal
   (c) Mechanical (d) Physical

102. **Minimum temperature for the pasteurization of milk is**
   (a) 125 °C (b) 90 °C
   (c) 71°C (d) 62°C

103. D-value is the time required to reduce the number of micro-organisms to:

(a) 10% (b) 90%

(c) 100% (d) 50%

104. Booster pump is used in HTST pasteurization to increase pressure of the milk in

(a) Heating section (b) Cooling section

(c) Holding section (d) Degeneration section

105. Pulsed electric field (PEF) uses

(a) Short electric pulses (b) Long electric pulses

(c) Both (a) & (b) (d) None of these

106. PEF stands for

(a) Pulsed electric field (b) Pulsed energy field

(c) Power electric field (d) Power energy filed

107. The main limitation of pulsed light technology

(a) Penetration depth (b) Heavy cost

(c) Side effects (d) None of these

108. Which are important mechanisms to explain heat generation in a material that is placed in a microwave field with increase in water content

(a) Ionic polarization (b) Dipole polarization

(c) Both (a) and (b) (d) None of these

109. Dielectric loss factor

(a) Increases (b) Decreases

(c) No change (d) Can't say

110. Full form of HHPP

(a) Hydrostatic high pressure processing

(b) High hydrostatic power processing

(c) High hydrostatic pressure processing

(d) High hydrodynamic power processing

111. PCM stands for

(a) Pest control management (b) Phase control mechanism

(c) Phase change materials (d) Phase change mechanism

112. NIR stands for

(a) Near infrared
(b) Near image resonance
(c) Nuclear infrared response
(d) Nuclear image response

113. NMRI stands for

(a) Nuclear magnetic response imaging
(b) Nuclear magnified resonance imaging
(c) Nuclear modified resonance imaging
(d) Nuclear magnetic resonance imaging

114. Vegetative microbial cells deactivated in HPP by

(a) Breaking non-covalent bonds
(b) Causing damage to cell membrane
(c) Both (a) and (b)
(d) None of these

115. The generation of pressure inside the vessel may be achieved by.

(a) Indirect compression
(b) Direct compression
(c) Heating of the pressure medium
(d) All of these

116. The critical electric field strength for plant cells

(a) 3-4 kV/cm
(b) 1-2 kV/cm
(c) 8-10 kV/cm
(d) >10 kV/cm

117. The critical electric field strength for microbial cells

(a) 8-10 kV/cm
(b) 14-16 kV/cm
(c) 10-12 kV/cm
(d) 10-14 kV/cm

118. The intensity of pulsed electric field varies from

(a) 10-20 Kv/cm
(b) 20-8O Kv/cm
(c) 80-100 Kv/cm
(d) None of these

119. In pulsed electric field, the cell membrane brake down

(a) Reversible
(b) Irreversible
(c) Both A &B
(d) None of these

120. Which type of food is treated in pulsed electric field Processing technique

(a) Solids
(b) Liquids
(c) Semi liquids
(d) Both (b) & (c)

121. The process of exposing food to gamma or x-rays is called

(a) Irradiation (b) HPP

(c) UV Technology (d) All the above

122. Irradiation sometimes called as

(a) Cold pasteurization (b) Hot pasteurization

(c) Both (a)&(b) (d) None of these

123. In oscillating magnetic field processing of food, the maximum frequency used

(a) 500 MHz (b) 600 MHz

(c) >800 MHz (d) 200-400 MHz

124. The technique used for measuring the dose known as

(a) Dosimetry (b) Radiometry

(c) Refractometery (d) Spectrophotometry

125. Low dose of irradiation used for

(a) Inhibition of germination, disinfection, ripening delay

(b) Shelflife extension, elimination of pathogens

(c) Sterilization

(d) Reduce or eliminate virus contamination

126. Medium dose of irradiation used for

(a) Reduce or eliminate virus contamination

(b) Sterilization

(c) Inhibition of germination, disinfection, ripening delay

(d) Shelf life extension, elimination of pathogens

127. 1 rad= Gy

(a) 0.01 (b) 0.0001

(c) 1 (d) 0.001

Answer Key

1.	(d)	21.	(c)	41.	(c)	61.	(c)	81.	(c)	101.	(a)	121.	(a)
2.	(b)	22.	(b)	42.	(d)	62.	(d)	82.	(b)	102.	(d)	122.	(a)
3.	(b)	23.	(b)	43.	(a)	63.	(c)	83.	(c)	103.	(b)	123.	(a)
4.	(a)	24.	(d)	44.	(b)	64.	(c)	84.	(a)	104.	(d)	124.	(a)
5.	(c)	25.	(a)	45.	(a)	65.	(a)	85.	(d)	105.	(a)	125.	(a)
6.	(a)	26.	(a)	46.	(a)	66.	(c)	86.	(c)	106.	(a)	126.	(d)
7.	(b)	27.	(c)	47.	(b)	67.	(b)	87.	(b)	107.	(a)	127.	(a)
8.	(c)	28.	(a)	48.	(a)	68.	(a)	88.	(d)	108.	(c)		
9.	(a)	29.	(b)	49.	(c)	69.	(c)	89.	(c)	109.	(a)		
10.	(c)	30.	(a)	50.	(b)	70.	(a)	90.	(b)	110.	(c)		
11.	(c)	31.	(d)	51.	(a)	71.	(c)	91.	(a)	111.	(c)		
12.	(b)	32.	(b)	52.	(b)	72.	(b)	92.	(c)	112.	(a)		
13.	(a)	33.	(c)	53.	(d)	73.	(d)	93.	(d)	113.	(d)		
14.	(c)	34.	(d)	54.	(d)	74.	(d)	94.	(a)	114.	(c)		
15.	(a)	35.	(c)	55.	(c)	75.	(c)	95.	(c)	115.	(d)		
16.	(b)	36.	(b)	56.	(a)	76.	(c)	96.	(a)	116.	(b)		
17.	(a)	37.	(c)	57.	(c)	77.	(b)	97.	(a)	117.	(d)		
18.	(a)	38.	(a)	58.	(b)	78.	(a)	98.	(c)	118.	(b)		
19.	(c)	39.	(b)	59.	(b)	79.	(b)	99.	(b)	119.	(b)		
20.	(c)	40.	(c)	60.	(a)	80.	(b)	100.	(d)	120.	(d)		

Chapter-8

Food Packaging

ONE LINER:

- The four Rs are: Reduce; Reuse; Recycle; Recover.
- The major groups are as follows- Polyolefins: PE and PP; Polyvinyl group: PVC; Condensation polymer: PET, PA; Styrene polymers: PS or Expanded PS (EPS); Carbonate group: PC; PAN: Polyacrilo nitrile.
- LDPE stands for low polyethylene; PP polypropylene; PVC: polyvinyl chloride; PS: polystyrene; PLA is Polylactic acid. PVdC: Polyvinyledene chloride.
- Densities of LDPE 0.910-0.925 g/cc; MDPE-0.926-0.94 g/cc andHDPE-0.94-0.965 g/cc.
- Heat resistances of LDPE 82-100°C; MDPE 105-121°C and HDPE-121°C.
- WVTR stands water vapor transmission rate& express in g/cm2/24 hr at 90% RH, 38°C.
- OTR stands for oxygen transfer rate.
- IIP stands for Institute of Institute of Packaging and its Headquarter located in Mumbai
- Carbon dioxide is mostly used in packaging.
- In Bread packaging, the low WVTR material is used.
- Fillers in papers improve gloss, brightness and opacity.

- CO_2 and O_2 is controlled in controlled atmospheric packaging (CAP).
- Risin is used in packaging to improve printability.
- PET is a used for soft drink packaging material.
- Light intensity and wavelength can have an affect on packed foods.
- In food packaging, aluminium foils are virtually impermeable to water vapour and gases.
- Tin cans range in thickness from 0.5 to 2 μ.
- Food is stored under lower pressure in a hypobaric storage.
- For strong acidic foods, I-type cans are used.
- The flex resistance test is used to determine how well a flexible packaging material seal.
- The oxygen barrier properties of ethylene vinyl alcohol are excellent.
- A polythene/packaging material should have bursting strength, tear resistant and grease resistant quality.
- First aseptic packaging of food was carried out by Neilson.
- Gas transmission rate (GTR) is measured in cc/24 hrs/m2/ atmp.@27°C
- Population strength of cannery wastes is ten times greater than domestic sewages.
- CFB stands for corrugated fibres boxes.
- Food packaging also aids in consumer communication, create brands, and facilitate marketing.
- A quaternary package is commonly used in interstate and international trade to assist the handling of tertiary shipments.
- Protection, containment, convenience and communication are all functions of packaging.
- The primary function of a package is to protect the commodity from the outside environment, mechanical forces, and other factors.
- The gas concentration is monitored continuously and chamber is impermeable in CAS.
- The rate of spoilage and discoloration is affected by both the intensity of light and the period of exposure.
- Sorption (Scalping) is the take-up of molecules from the food by the package e.g. scalping of flavor compounds from fruit juices by plastic.

- The transfer of molecules from the packing material into the product is known as migration.
- Plasticizers, residual monomers, and antioxidants transfer from packaging materials to food and, in certain cases, the environment.
- LDPE used as a film wrapping while HDPE used in both film form and rigid packs.
- HDPE can be used in both form and as rigid packs suchfilm asbottle form.
- Sheet extrusion is an application of profile extrusion, but it uses a die with a slot orifice. Film product formed by this process is referred to as "cast film".
- Tin-free steel (TFS) is also referred to as electrolytic chromium/ chrome oxide coated steel (ECCS).
- A Stay-on Tab (SOT) enables the opening tab and pierce-open end section to be retained on the can.
- Impact extrusion is another method by which ductile metals such as tin, lead and aluminum can be formed to tubular shapes.
- Emerald glass is a bright green glass used mostly for wines and lime or lemon flavored soft drinks.
- Amber (brown) glass is the only standard glass that will filter out light in the critical ultraviolet (UV) region (300 to 400 nm) and used for UV-sensitive products such as beer and some pharmaceuticals.
- The mechanical shears (which are 25 mm below the orifice and synchronized with the draw-off flow rate and bottle-forming machine speed) snip off "gobs" of molten glass.
- Glass manufacturing method is 'press and blow'.
- As per US practices, paper board constitutes materials more than 300 μm thick.
- Evenness of fiber distribution is referred to as "formation".
- The chemical pulping can be further classified as alkali sulphate and acid sulphite methods.
- Mechanical pulping involves grinding wood.
- Grease resistant papers are made from highly refined pulps and high refining levels also reduce paper opacity.
- Sheet forming normally takes place on a wire mesh covered cylinder.

Questions

1. **A secondary package, such as package, is made up of many stretch-wrapped pallets of corrugated cases.**

   (a) Quaternary (b) Secondary

   (c) Tertiary (d) None of the above

2. **After sheet formation, the sheet is pressed to reduce the moisture content to .... absorbent blankets and gently pressing by steel rolls.**

   (a) 70-75% (b) 60-65%

   (c) 40-45% (d) 10%

3. **Food packaging serves which purposes?**

   (a) Convenience (b) Containment

   (c) Protection (d) All of these

4. **The main focus, in the selection of suitable packaging materials for a particular food?**

   (a) Aesthetic appeal (b) Cost

   (c) Barrier properties (d) Availability of the packaging material.

5. **The shelf life of a food is altered by a change in**

   (a) The composition and formulation

   (b) Processing parameters and Packaging system

   (c) The environment to which it is exposed

   (d) All of the these

6. **Packaging can be used to create a barrier against..........agents that cause food to spoilage.**

   (a) Both biotic and abiotic (b) Biotic

   (c) Abiotic (d) Neither biotic nor abiotic.

7. **Food deterioration is mostly caused by chemical reactions in the food?**
   (a) Oxidation reactions
   (b) Neither oxidation nor enzymatic browning reactions
   (c) Non enzymatic browning reactions
   (d) Either oxidation or enzymatic browning reactions

8. **On packaged food, a green coloured square or triangle with a green dot in the centre denote that the food is**
   (a) Vegetarian (b) Processed by irradiation
   (c) Treated with preservatives
   (d) Non-meat food.

9. **The basis of...... is to flush the inside of a permeable package with a gas such as CO_2 or N2 before sealing it.**
   (a) Controlled atmosphere packaging (CAP)
   (b) Modified atmosphere packaging (MAP)
   (c) Neither MAP nor CAP
   (d) Both MAP and CAP

10. **Catalytic effects of light are most pronounced in the ....wavelengths of the UV spectrum and visible spectrum.**
   (a) Higher (b) Medium
   (c) Lower (d) all

11. **Food packaging polymers can be modified to prevent light from entering the package via...........**
   (a) Either incorporation of dyes or application of coatings
   (b) Application of coatings
   (c) Incorporation of dyes
   (d) None of the above

12. **.............. chemical is used as sterilizing medium in aseptic-packaging?**
   (a) H_2SO_4 (b) H_2O_2
   (c) HCI (d) CH_3COOH

13. **In addition to permeability, mass transport phenomena in package systems are,**
   (a) Sorption and migration (b) Sorption
   (c) Migration (d) None of above

14 **.................. packaging materials has the potential to change the moisture content of food during long-term storage?**

(a) Glass (b) Paper

(c) PET (d) None of the above

15. **The thickness of aluminum foil is**

(a) 1 mm (b) 1-3 μm

(c) 4-150 μm (d) 4-150 mm

16. **The rule is to sterilise both the foods and the packing materials separately, then assemble them in a sterile atmosphere.**

(a) Smart packaging (b) Consumer packaging

(c) Aseptic packaging (d) Retort packaging

17. **Because of this............., some wines and spirits continue to use wood.**

(a) Transfer of flavor compounds from wood

(b) Good stacking characteristics

(c) High weight-to-strength ratio

(d) Good mechanical protection

18. **Chemically treated woven jute sacks**

(a) To prevent rotting

(b) To reduce their flammability

(c) To make it moisture proof

(d) Both to prevent rotting and to reduce their flammability

19. **In early 1990s, the guiding principles for designing environmentally responsible packaging developed is referred to as .... hierarchy.**

(a) four Rs (b) four Ps

(c) four Ms (d) four Es

20. **A thermo set plastic is ?**

(a) Electrical switch (b) Plastic carry bag

(c) PET bottles (d) Polyster

21. **The PE and PP belong to a broader group which is called as**

(a) ployolefins (b) condensation polymers

(c) carbonate group (d) styrene polymers

22. HDPE is commonly recycled, and has the identification code number........in its recycling symbol.

(a) 1 (b) 2

(c) 3 (d) 2

23. The resin identification codes which are given to help in recycling have been given by

(a) Association of Plastics Manufacturers in Europe

(b) Society of Plastics Industry

(c) American society of Plastic manufacturer and recyclers

(d) Indian Institute of Packaging

24. .... is also known as Nylon?

(a) Polyamide (b) PVC

(c) Polystyrene (d) PLA

25. Teflon is

(a) Polytetrafluoroethylene (PTFE/PCTFE)

(b) Polycarbonate

(c) PVdC

(d) PVC

26. Saran is a trade name of

(a) PVC (b) PVdC

(c) PET (d) PS

27. .... is a biodegradable plastic?

(a) PET (b) PLA

(c) PAN (d) PCTFE

28. .... is known as thermocole?

(a) Polystyrene (PS)

(b) Expanded polystyrene (EPS)

(c) Poly (ethylene terephthalate) glycol (PETG)

(d) Polyvinyl acetate (PVA)

29. The difference between plastic films and sheets is that the plastic films are less than ......

(a) 60 um (b) 75 m

(c) 100 um (d) 150 um

30. **Which of the following is commonly used in plastic bags?**

(a) LDPE (b) HDPE

(c) PP (d) PET

31. **High density PE is used for**

(a) Film wrapping (b) boil-in-the-bag products

(c) store oily foods (d) None of the above

32. **Milk bottles and shampoo bottles are made up of**

(a) LDPE (b) PP

(c) HDPE (d) PET

33. **The water bottles, soft drink bottles are made from**

(a) PP (b) PET

(c) PS (d) PLA

34. **The electrical conduits as well as plumbing pipes, garden hose are made from**

(a) HDPE (b) PVOH

(c) mPP (d) PVC

35. **The CD cases, yoghurt pots are examples of**

(a) PP (b) PS

(c) EPS (d) PVdC

36. **............ can be used for shrink wrapping?**

(a) PP film (b) HDPE

(c) PVC film (d) PS film

37. **The individual fruits are shrink wrapped in a tunnel,in which the temperature and time maintained are**

(a) 80-100°C for 15-30 s (b) 180-210°C for 15-30 s

(c) 80-100°C for 2-3 min (d) 180-210°C for 2-3 min

38. **LDPE contains the chemical elements..........**

(a) Carbon and hydrogen

(b) Carbon, hydrogen and oxygen

(c) Carbon, hydrogen and nitrogen

(d) Carbon, hydrogen, oxygen and nitrogen

39. **.................. film stretches maximum?**

(a) Low density polyethylene (b) polypropylene

(c) Cellulose (d) high density polythene

40. LDPE has ....branching than HDPE.

(a) Same (b) less

(c) More (d) None

41. The mass density of high-density polyethylene usually ranges from

(a) 0.93 to 0.97 g/cm^3 (b) 0.53 to 0.65 g/cm^3

(c) 1.23 to 1.50 g/cm^3 (d) 2.35 to 3.00 g/cm^3

42. The LDPE has a maximum heat resistance of

(a) 60°C (b) 80°C

(c) 100°C (d) 120°C

43. RFID stands for

(a) Radio frequency identification

(b) Radio frequency indicator

(c) Radio frequency identity detection

(d) Radio frequency infrared detection

44. Which of the following packaging material is a good moisture barrier, but has relatively high gas permeability

(a) Polypropylene (b) LDPE

(c) Plain cellulose (d) PET

45. The outstanding properties of polyethylene terephthalate (PET) film as a food packaging material are due to its

(a) Great tensile strength (b) Excellent chemical resistance

(c) Light weight, elasticity (d) All of the above

46. Which of the following has an excellent barrierproperty, and hence used as bottles for soft drinks?

(a) LDP (b) HDPE

(c) PET (d) PVC

47. In plastic manufacturing, a shape of constant cross-section can be extruded by forcing polymer melt through a shaped orifice in a die placed at the exit port. This is known as

(a) Cast film (b) Blown film

(c) Profile extrusion (d) Blow molding

48. The hollow plastic parts are made by a process knownas:

(a) Profile extrusion (b) Cast film extrusion

(c) Blown film extrusion (d) Blow molding

49. Aluminum foil has excellent barrier properties, provided it is at least ..... thick.

(a) 10 μm (b) 15 μm

(c) 20 μm (d) 25 μm

50. The steel used for metal cans are usually

(a) Low carbon steel (b) Medium carbon steel

(c) High carbon steel (d) Stainless steel

51. The term tinplate refers to................

(a) Thin plate made of tin

(b) Low carbon mild steel sheet with a coating of tin on each surface of the material.

(c) Low carbon mild steel sheet with a coating of tin on one side of the material.

(d) None of the above

52. During can making, the coatings and inks are normally dried by passing the sheets through a heate oven where the temperature is about

(a) 60°C (b) 120°C

(c) 180°C (d) more than 300°C

53. In can making, DRD can' is a term used for

(a) Double rolled ductile can (b) Double resistance drawn can.

(c) Drawn and rolled ductile can (d) Drawn and redrawn can.

54. In can making, 'DWI is a term used for

(a) Double walled and ironed (b) Drawn and wall ironed

(c) Drawn and wall insulated (d) None of the above.

55. In DWI cans, ironing operation is done to

(a) Make the wall smooth by pressing

(b) Make the wall corrosion resistant by a layer lacquering.

(c) Make the wall thin by passing through hardened circular dies.

(d) Have a better fit of the covers.

56. The DWI cans for processed food are made from only tin plate, not from aluminium because

(a) thin wall aluminit cans cannot provide 100% freedom from pin holes

(b) thin wall aluminium cans do not have sufficient strength to withstandt the heat process cycles. Sufficient

(c) aluminium cans develop off flavour during storage of processed foods

(d) Aluminium cans are deformed easily during subsequent handling.

57. For drink cans, the common form of top is

(a) Stay on Tab (SOT) (b) Roll on

(c) Full aperture easy open (d) Press on / twist off metal closure

58. Which of the following is not related to the making of a two piece can?

(a) Draw (b) Draw and redraw

(c) Draw and wall ironing (d) Impact extrusion

59. Double seaming is a method of

(a) Making the canner's end of a can

(b) Making the cylindrical shape of the can

(c) Rolling the steel to make it thin

(d) None of the above

60. Which of the following containers have full aperture easy-open ends (FAEOs)?

(a) Metal (b) Plastic

(c) Glass (d) Paper

61. The basic clear glass used for majority of packaging applications is known as

(a) Flint (b) Amber

(c) Emerald (d) None of the above

62. The beer bottles are generally made of ....glass.

(a) Amber (b) Blue

(c) Green (d) Flint

63. For preparation of amber glass, which of the following is added?

(a) Chrome oxide (b) iron and sulfur

(c) cobalt oxides (d) iron oxides

64. **In the glass manufacturing process, the raw materials fuse into glass at about**
    (a) 800°C (b) 1200°C
    (c) 1500°C (d) 2100°C

65. **The molten glass is allowed to flow through a temperature controlled channel known as forehearth', in which the temperature is dropped down to**
    (a) 800°C (b) 1000°C
    (c) 1200°C (d) 1300°C

66. **In the glass manufacturing process; the parison is prepared in the**
    (a) Blank mold (b) Blow mold
    (c) IS machine (d) Forehearth

67. **................. is a glass manufacturing process?**
    (a) Blow and blow (b) Profile extrusion
    (c) Blow molding (d) Drawing and redrawing

68. **After the bottles are prepared, the bottles are slowly cooled which is known as**
    (a) Cold end surface treatment (b) Annealing
    (c) Second surface treatment (d) Tempering.

69. **The main constituent of wood that is important in paper making is**
    (a) Starch (b) Cellulolse
    (c) Pentosan (d) Fibres

70. **As per the ISO, the paper board constitutes material weighing**
    (a) More than 250 g/m^2 (b) More than 100 g/m^2
    (c) More than 200 μm thick (d) More than 300 μm thick.

71. **Which of the following is better for short fibre length materials than the long ones?**
    (a) Tear strength (b) Burst strength
    (c) Tensile strength (d) Formation

72. **The paper quality depends on the length of fibres obtained from the woods. The long fibres have the average fibre length of**
    (a) 1-1.5 mm (b) 2-3 mm
    (c) 3-4.5 mm (d) 6-7 mm

73. The paper quality depends on the length of fibres obtained from a woods. The short fibres have the length of

(a) 1-1.5 mm (b) 2-3 mm

(c) 3-4.5 mm (d) 6-7 mm

74. Pulping is the process of

(a) Extracting virgin or primary fibre directly from wood

(b) Adding chemicals to improve the quality of fibres.

(c) Adding bleaching earth to the fibres

(d) Adding alkali to improve the property.

75. .... is commonly prepared from the mechanical pulping process?

(a) Kraft bags (b) Newsprint

(c) Glossy paper (d) Thick paper

76. The kraft paper are produced by

(a) Mechanical pulping

(b) Alkali sulfate pulping

(c) Acid sulfites pulping

(d) Laminating paper with bonding agents.

77. Glassine is a paper which has/is

(a) Maximum fibril separation.

(b) Prepared from highly refined pulps.

(c) Bleached chemically with chlorine containing compounds or hydrogen peroxide.

(d) A separate coating has been applied on the paper.

78. Increased refining of the pulp increases

(a) Absorbency (b) Tear strength

(c) Opacity (d) Tensile strength

79. During stock preparation, the fibre in water suspension has about.....% fibre and ....% water.

(a) 2, 98 (b) 10, 90

(c) 20, 80 (d) 40, 60

80. Alums and synthetic resins are added during stock preparation to

(a) Increase water repellency of fibre

(b) Increase color

(c) Increase strength

(d) Increase smoothness

81. .... gas composition is used in MAP,

(a) O_2, N_2 (b) CO_2, N_2,

(c) CO_2, O_2 (d) CO_2, O_2, N_2

82. White board is suitable for contact with food and is often coated with polyethylene, PVC or wax for ....

(a) Greater mechanical strength

(b) Heat sealability

(c) Both improved appearance and strength

(d) Better printing

83. The Chip board is often lined with white board for

(a) Greater mechanical strength

(b) Heat sealability.

(c) Both Improved appearance and strength

(d) Better printing.

84. Paper board used for food packaging is often coated with a wax or polythene mainly to,...............

(a) Prevent interaction with contents

(b) Prevent oxygen permeability

(c) Prevent moisture permeability

(d) Prevent oil migration

85. The creation of an appropriate gaseous atmosphere around a commodity in selective permeable polymeric film packages to enhance shelf life is known as

(a) CAS (b) MAP

(c) Shrink packaging (d) Flexible film packaging

86. In a MAP, if the atmosphere is established by pulling a slight vacuum and replacing the package atmosphere with the desired gas mixture, then it is known as

(a) Intelligent packaging (b) Active modification

(c) Passive modification (d) None of the above

87. In MAP, if the modification of atmosphere is attained only through respiration of commodity within package and depends on characteristics of commodity and packaging film only, and there is no control on initial gas composition, then the type of MAP is known as

(a) Intelligent packaging (b) Active modification

(c) Passive modification (d) None of above

88. ....is packaging in which subsidiary constituents have been deliberately included in or on either the packaging material or the package headspace to enhance the performance of the package system.

(a) Active packaging (b) Active modification

(c) Passive modification (d) Smart packaging

89. .... is an example of smart packaging?

(a) Oxygen absorber (b) Time temperature indicator

(c) Self heating/ cooling (d) Antimicrobial packaging film

90. For which of the following type of food, MAP cannot be practised?

(a) Fresh fruit and vegetables (b) Meat product

(c) Bakery products (d) None of the above

91. In a typical CA storage, oxygen is decreased from the normal 21 o in air to:

(a) 1 to 3% (b) 3 to 5%

(c) 10-12% (d) None of the above

92. In a typical CA storage, carbon dioxide is increased from 0.04 % level to:

(a) 0.5-2% (b) 2-5%

(c) 10-12% (d) None of the above

93. CA storage of apple is normally done at

(a) 4-5°C, 2-3% oxygen and 5-6% CO_2

(b) 0-5°C, 2-3% oxygen and 1-2% CO_2

(c) 0-5°C, 7-8% oxygen and 1-2% CO_2

(d) None of the above

94. .... is a intelligent packaging system

(a) Oxygen scavengers (b) Time temperature indicator

(c) CO_2, emitters (d) Antimicrobial packaging

95. In an actively modified MAP/CAS, which of the following can be used as an oxygen absorber?

(a) Ferrous oxide (b) Magnesium oxide

(c) Activated charcoal (d) Potassium permanganate

96. Besides the iron powders, which of the following can also act as oxygen scavenger in active packaging systems?

(a) Ascorbic acid (b) Alcohol oxidase

(c) Glucose oxidase (d) All of the above

97. Sachets that contain. .... can absorb CO_2.

(a) $Ca(OH)_2$ (b) $Fe(OH)_2$

(c) Potassium permanganate (d) None of the above

98. In an actively modified MAP/CAS,...............can be used as a CO_2 absorber?

(a) Ferrous oxide (b) Cobalt oxide

(c) Activated charcoal (d) Potassium permanganate

99. In an actively modified MAP/active packaging....can be used as an ethylene absorber?

(a) ferrous Oxide (b) magnesium oxide

(c) calcium oxide (d) potassium permanganate

100. In a CA storage, dry ice is put in the storage to

(a) Absorb oxygen (b) Reduce CO_2% level

(c) Increase CO_2, level (d) None of the above

101. Dry ice is

(a) Solid water kept in moisture free environment

(b) Solidified freon.

(c) Solidified methane

(d) Solidified CO_2

102. Which of the following can be used for moisture absorption in dry food packages

(a) Calcium oxide (b) Magnesium oxide

(c) Sulphur dioxide (d) None of the above

103. Which of the following can be used to release antioxidant into a package?

(a) Nisin (b) α-tocopherol

(c) Ascorbic acid (d) $Ca(OH)_2$

104. The self heating cans use a reaction of

(a) Ascorbic acid and lime. (b) Lime and water

(c) CaSO, and water (d) Dil HaSO, and chlorine.

105. The self cooling cans involve endothermic dissolutionof

(a) Ammonium nitrate and chloride in water

(b) $CuSO_4$ in water.

(c) Citric acid in water

(d) None of the above

106. ....stored produce deteriorate rapidly when exposed to normal atmospheres during marketing,

(a) Controlled atmosphere (b) Modified atmosphere

(c) Active packaging (d) Smart packaging

107. The first aseptic packaging of food was carried out for

(a) Milk in paper board carton (b) Fruit juice in paper board carton

(c) Milk in metal cans (d) Fruit juice in laminated packages

108. Which surface of the carton material in aseptic packaging paperboard laminate carton is coated with polyethylene?

(a) Internal (b) External

(c) Both internal and external (d) Either internal or external

109. ....instrument is used to measure the thickness of a packaging material?

(a) Dial gauge &screw gauge (b) Micrometer

(c) Vernier calipers (d) All of the above

110. .... test measures the ability of the packaging film to withstand pneumatic or hydraulic pressure built up?

(a) Burst strength (b) Tensile strength

(c) Impact strength (d) All of the above

111. .....tests are designed to measure the ability of the films to withstand fracture by shock?

(a) Burst strength (b) Tensile strength

(c) Impact strength (d) All of the above

112. Which of the following property of the packaging film is expressed with the unit of "grammage"?

(a) Paper density (b) Paper mass

(c) Paper thickness (d) All of the above

113. The gram of water vapour transmitted from 1 m^2 if film area in 24 hours is known as

(a) Water vapour permeance

(b) Water vapour transmission rate(WVTR)

(c) Water vapour permeability

(d) Moisture transfer coefficient

114. The gram of water vapour transmitted from I m^2 if film area in 24 hours when the vapour pressure difference of 1 mm of Hg is called

(a) Water vapour permeance

(b) Water vapour transmission rate(WVTR)

(c) Water vapour permeability

(d) Moisture transfer coefticient

115. The gram of water vapour transmitted from 1 m^2 if film area in 24 hours when the vapour pressure difference of 1 mm of Hg and 1 cm of film thickness is termed as

(a) Water vapour permeance

(b) Water vapour transmission rate(WVTR)

(c) Water vapour permeability

(d) Moisture transfer coefficient

116. What are the roles of packaging materials?

(a) Protection

(b) Communication

(c) Convenience and Containment

(d) All of the above

117. What is the primary function of packaging material?

(a) Containment (b) Protection

(c) Communication (d) convenience

118. Non crystalline of polymers are characterized by

(a) Glass transition temperature (b) Boiling temperature

(c) Melting temperature (d) All of the above

119. Crystalline polymers are characterized by

(a) Boiling temperature (b) Glass transition temperature

(c) Both a and b (d) Crystalline melting temperature

120. The relation between Tg and Tm for crystalline unsymmetrical materials are

(a) Tg=Tm (b) Tg-2/3Tm

(c) Tg-3Tm (d) Tg=1/2Tm

121. The relation between Tg and Tm tor crystalline symmetrical materials are

(a) Tg-3Tm (b) Tg-Tm

(c) Tg-1/2Tm (d) Tg-2/3Tm

122. The density of HDPE films are

(a) >=940kg/m3 (b) <550 kg/m3

(c) <940 kg/m3 (d) >550 kg/m3

123. The density of LDPE films are

(a) 315-439 kg/m3 (b) 915-939 kg/m3

(c) 995-1039 kg/m3 (d) 615-839 kg/m3

124. The crystallinity of LDPE varies from

(a) 60-80% (b) 30-40%

(c) 70-90% (d) 50-70%

125. The softening point of LDPE lies

(a) Just below 100°C (b) >150°C

(c) >100°C (d) >170°C

126. ESCR stands for

(a) Electrical strength crack resistance

(b) Electrical stress check resistance

(c) Electrical stress crack resistance

(d) Electrical strength check resistance

127. The formation of sheet metal without changing its thickness is known as:

(a) Beating (b) Drawing

(c) Softening (d) All of the above

128. The operation of reforming two piece of can into one is known as

(a) Re-drawing (b) Soldering

(c) Welding (d) All of the above

129. Drawn and re-drawn. Containers are often referred as

(a) DR cans (b) DRD cans

(c) DDR cans (d) None of these

130. The process of thinning the wall of two-niece can by passing it through hardened circular dies

(a) Beating (b) Soldering

(c) Welding (d) Ironing

131. DWI stands for

(a) Draw and wait ironing process

(b) Draw and worm ironing process

(c) Draw and wall ironing process

(d) Draw and welding ironing process

132. Glass is made up of

(a) Silicate, magnesium and soda (b) Silicates, lime and soda

(c) Calcium, magnesium and soda (d) Silicate, magnesium and iron

133. Clear glass is also known as

(a) Black flint (b) Clear flint

(c) Light flint (d) White flint

134. Pale green glass can be obtained by rising the ....content

(a) Iron (b) Ca

(c) Mg (d) Cu

135. The dark green colour of glass is obtained by adding

(a) Magnesium oxide and calcium oxide

(b) Chromium oxide and iron oxide

(c) Copper oxide and calcium oxide

(d) Iron oxide and calcium oxide

136. .... glass is used for light-sensitive products

(a) Clear glass (b) Pale green glass

(c) Amber glass (d) All of the above

137. Blue glass can be obtained by addition of

(a) Cobalt to a low-iron glass (b) Calcium to a low-iron glass

(c) Iron to a low-iron glass (d) Magnesium to a low-iron glass

138. .... characteristic is not related to glass

(a) Less strength
(b) Chemically inert
(c) Transparent
(d) Odour less

139. .... combination is correct for hot filled products?

(a) Filled at 5°C and headspace 85%
(b) Filled at 55°C and headspace 5%
(c) Filled at 75°C and headspace 15%
(d) Filled at 85°C and headspace 5%

140. Which of the following combination is correct for cold filled products?

(a) Sterilizing at 131°C and head space 16% minimum
(b) Sterilizing at 71°C and head space 6% minimum
(c) Sterilizing at 121°C and head space 6% minimum
(d) Sterilizing at 131°C and head space 6% minimum

141. .... polyethylene material is used in shrink wrap packaging

(a) VLDPE
(b) PET
(c) HDPE which of the following is not a property of LDPE
(d) PVdC

142. .... is a property of HDPE

(a) High softening and melting points
(b) Good moisture barrier and poor oxygen barrier
(c) High cost
(d) Difficult to processed

143. .... is incorrect about polypropylene

(a) UV resistant
(b) Very good chemical resistance
(c) Excellent moisture resistance
(d) Low coefficient of friction

144. Calendering is a process of

(a) Making the paper in to rolls
(b) Squeezing the sheet through a series of steel/compression rolls to enhance smoothness and to get uniform thickness.
(c) Applying a layer of coat and then pressing to make
(d) Controlled drying to improve the strength of the paper

145. Meting point of PET is

(a) 360°C (b) 160°C

(c) 460°C (d) 260°C

146. Which of the following is incorrect about polyamide

(a) High permeable to moisture and vapour

(b) Good flavor and odour barrier

(c) Not heat sealable

(d) Not resistant to oil and fat

147. Which of the following is not a property of PVdC

(a) Less barriers to gases (b) Barrier to fats and oils

(c) Barrier to water vapour (d) Heat sealable

148. To avoid thermal shock, cool down differential should not exceed

(a) 30°C (b) 40°C

(c) 50°C (d) 60°C

149. To avoid thermal shock, warm up differential should not exceed

(a) 65°C (b) 45°C

(c) 75°C (d) 55°C

150. The amount of fibre in paper is expressed as

(a) Weight per unit area (b) Appearance

(c) GSM, Thickness (d) All of the above

151. The paper over ....g/m² is defined by ISO as paper board

(a) 300 (b) 200

(c) 400 (d) 500

152. .... is not a property of a paper?

(a) Permeable to water (b) Heat sealable

(c) Permeable to gases (d) Less mechanical strength

153. .... method is used to impart barrier properties of paper

(a) Hard sizing (b) Lamination

(c) Sizing with wax (d) All of the above

154. Blown film are

(a) Tough (b) Resilient

(c) Both (a) and (b) (d) None of these

155. Cast film are

(a) Soft (b) Easy to stretch

(c) None of these (d) Both (a) and (b)

156. The lower mechanical properties of cast film are due to

(a) Fast cooling

(b) Single orientation of molecular structure

(c) Both a and b

(d) None

157. The higher mechanical properties of blown film are due to

(a) Slow cooling (b) Compositeorientationofthemolecular structure

(c) None of these (d) Both (a) and (b)

158. .... is drawback of biopolymer

(a) Poor water vapour barriers (b) Poor mechanical properties

(c) Both A and B (d) None

159. The gas barrier properties of biopolymers depend on the

(a) Ambient temperature (b) Ambient humidity

(c) Ambient pressure (d) All of the above

160. .... is not active packaging system?

(a) Time temperature indicator (b) RFIDD

(c) Integrity indicator (d) All of the above

Answer Key

1.	(c)	21.	(a)	41.	(a)	61.	(a)	81.	(d)	101.	(d)	121.	(c)	141.	(a)
2.	(b)	22.	(b)	42.	(c)	62.	(a)	82.	(c)	102.	(a)	122.	(a)	142.	(b)
3.	(d)	23.	(b)	43.	(a)	63.	(b)	83.	(c)	103.	(b)	123.	(b)	143.	(a)
4.	(c)	24.	(a)	44.	(b)	64.	(c)	84.	(a)	104.	(b)	124.	(d)	144.	(b)
5.	(d)	25.	(a)	45.	(d)	65.	(d)	85.	(b)	105.	(a)	125.	(a)	145.	(d)
6.	(a)	26.	(b)	46.	(c)	66.	(a)	86.	(b)	106.	(a)	126.	(c)	146.	(d)
7.	(d)	27.	(b)	47.	(c)	67.	(a)	87.	(c)	107.	(c)	127.	(b)	147.	(a)
8.	(a)	28.	(b)	48.	(d)	68.	(b)	88.	(a)	108.	(c)	128.	(a)	148.	(b)
9.	(b)	29.	(c)	49.	(d)	69.	(b)	89.	(b)	109.	(d)	129.	(b)	149.	(a)
10.	(c)	30	(a)	50.	(a)	70.	(a)	90.	(d)	110.	(a)	130.	(d)	150.	(d)
11.	(a)	31.	(b)	51.	(b)	71.	(d)	91.	(b)	111.	(c)	131.	(c)	151.	(b)
12.	(b)	32.	(c)	52.	(c)	72.	(c)	92.	(b)	112.	(b)	132.	(b)	152.	(b)
13.	(a)	33.	(b)	53.	(d)	73	(a)	93.	(b)	113.	(b)	133.	(d)	153.	(d)
14.	(b)	34.	(d)	54.	(b)	74.	(a)	94.	(b)	114.	(a)	134.	(a)	154.	(c)
15.	(c)	35.	(b)	55.	(c)	75.	(b)	95.	(a)	115.	(c)	135.	(b)	155.	(d)
16.	(c)	36.	(c)	56.	(d)	76.	(b)	96.	(d)	116.	(d)	136.	(c)	156.	(c)
17.	(a)	37.	(b)	57.	(a)	77.	(a)	97.	(a)	117.	(b)	137.	(a)	157.	(d)
18.	(d)	38.	(a)	58.	(d)	78.	(d)	98.	(c)	118.	(a)	138.	(a)	158.	(a)
19.	(a)	39.	(a)	59.	(a)	79.	(a)	99.	(d)	119.	(d)	139.	(d)	159.	(b)
20.	(a)	40.	(c)	60.	(a)	80.	(a)	100.	(c)	120.	(b)	140.	(c)	160.	(d)

Chapter-9

Dairy Engineering

ONE LINER:

- Flow diversion valve is used to divert un-pasteurized milk to the balance tank.
- With the increase in process temperature, the thermal death time decrease.
- Main objective of pasteurization is to destroy all harmful mirco-organism.
- The Z value is the temperature required to raise so that decimal reduction time is reduced by 90%.
- Booster pump is used in HTST pasteurization to increase pressure of the milk in regeneration section.
- In thermal processing, decimal reduction time refers to the time required to destroy 90% of micro- organisms.
- Direct method of heating for UHT processing is useful for low viscosity liquid.
- The quality of steam increased with an increase in the total heat content of the steam above a threshold temperature.
- The zone of slowest heating is called the Cold point of a container.
- Ten fold changes in TDT is known as Z value.
- In HTST pasteurizer, milk is heated for 15 seconds at 72-75 temperature°C.

- Blanching is primarily carried out to inactivate enzymes.
- Pasteurization is a process in which harmful microorganism are destroyed
- The number of times a reaction rate changes with a 10°C change in temperature is called Q10 value.
- The time required for achieving a stated reduction in the microbial population at a given temperature, is called decimal reduction time.
- The fluid milk is pasteurized at 150 °C for 1 second to improve its durability and to kill disease causing micro-organisms.
- In UHT sterilization, milk is sterilized in a continuous flow sterilizer at temperature of130-150 °Cfor few second.
- Dry heat is less sporicidal as compared to the moist heat at same temperature.
- Two factors that affect heat requirement for pasteurization are temperature and time.
- Blanching is used for deactivating enzymatic reaction.
- Thermal death time decreases as temperature increase.
- The pasteurizing process was developed by Louis Pasteur.
- Pasteurization of milk destroys harmful micro-organism.
- The usual range of temperature and time employed in UHT process is 130-150 °C and 1-20 seconds.
- The usual of range of temperature and time employed in sterilization process is 140 °C and 3-4 seconds.
- Time-temperature combination for HTST pasteurization is 72-75 °Cand 15 sec.
- UHT processing is also known as aseptic processing.
- HTST pasteurization system adapts itself well to CIP cleaning.
- D and Z values for destruction of nutritional and quality attributes are larger than that of the microorganisms.
- During pasteurization, temperature is maintained below 100°C.
- Ultra high temperature (UHT) type processing has very effective sterilization effect.
- A quotient Q10indicates the change in reaction rate for 10°C change in temperature.
- The main purpose of pasteurization is to destroy pathogenic organisms.

- ❒ Hydrostatic sterilizer is used for sterilization of canned food.
- ❒ UHT is used in aseptie processing.
- ❒ Microbial destruction is' a first order reaction.
- ❒ Spore forming bacteria are more heat resistant than vegetative bacteria.
- ❒ Pasteurization does not kill all micro-organisms.
- ❒ The HTST pasteurization system adopts itself well to CIP system.
- ❒ The F value is defined as the number of minutes required to destroy a stated number oforganisms at a defined temperature usually 121°C.
- ❒ Bottle sterilization of milk is at 105-110°C for 30-45 minute.
- ❒ High acid foods need less heat treatment than low acid foods.

Questions

1. .... milk is rich in appearance?

(a) Pasteurized milk (b) Irradiated milk
(c) Sterilized milk (d) Homogenized milk

2. Which of the following is the quickest in digestion by the infants.

(a) Homogenized milk (b) Pasteurized milk
(c) Boiled milk (d) Mother's milk

3. A quart bottle is taken and the visible cream separation is observed to identify

(a) Pasteurized milk (b) Sterilized milk
(c) Homogenized milk (d) Butter milk

4. ....is recommended to avoid cream plug formation of milk?

(a) Pasteurization (b) Homogenization
(c) Boiling (d) Churning

5. **Which of the following operation increases the stability of the emulsion, so that the cream does not separate on standing?**

   (a) Sterilization (b) Homogenization

   (c) Pasteurization (d) Preservation

6. **In case of irradiated milk, which of the following vitamin is normally increased?**

   (a) Vitamin A (b) Vitamin C

   (c) Vitamin B (d) Vitamin D

7. **The milk is ....to increase the keeping quality so that it does not spoil during distribution.**

   (a) Homogenized (b) Pasteurized

   (c) Defatted (d) None of the above

8. **Milk is an excellent source of minerals..........**

   (a) Calcium and phosphorus (b) Magnesium and iron

   (c) Sodium and potassium (d) None

9. **The milk to which vitamins and minerals are added is known as**

   (a) Fortified milk (b) Filled milk

   (c) Recombined milk (d) All the above

10. **...................... is common in milk?**

   (a) Acetic acid fermentation (b) Bacterial fermentation

   (c) Lactic acid fermentation (d) None of the above

11. **Lactic acid provides acidity in the intestinal tract,which facilitates absorption of.......And..............**

   (a) Calcium, phosphorus (b) Calcium and magnesium

   (c) Iron and calcium (d) Phosphorus and iron

12. **Milk provides a substrate for lactic acid bacteria, and tends to suppress ...in the intestine.**

   (a) Fermentative bacteria (b) Putrefactive moulds

   (c) Putrefactive bacteria (d) Fermentative moulds

13. **.................... is obtained as a by-product of butter making?**

   (a) Natural butter milk (b) Cultured butter milk

   (c) Acidophilus milk (d) None of the above

14. Cow's milk contains about ....% fat.

(a) 3 (b) 7

(c) 4 (d) 10

15. Human milk contains about .... % fat?

(a) 4 (b) 3

(c) 15 (d) 9

16. The fat content of yoghurt varies from

(a) 0-5% (b) 5-15%

(c) 15-25% (d) 25-45%

17. Preparation of curd is one method of prolonging the shelf-life of milk by several days. Soured milk does not readily undergo ....

(a) Lipolysis (b) proteolysis

(c) hydrolysis (d) none of above

18. Biochemical reaction of proteolysis by the enzyme proteases results in formation of......

(a) amino acids and peptides (b) bitter taste

(c) changes in texture (d) all of these

19. The ..... and ....content of milk increase during curd formation.

(a) vitamin A and vitamin C (b) riboflavin, thiamine

(c) vitamin D and vitamin A (d) None of above

20. Paneer contains about.......... per cent protein.

(a) 18 (b) 38

(c) 48 (d) 8

21. Cheese contains about .... per cent protein.

(a) 14 (b) 44

(c) 24 (d) 54

22. The main proteins in milk are ....

(a) Glycoprotein (b) Lipoprotein

(c) casein and lactalbumin (d) All of these

23. Which of the following acids are commonly used for preparation of chhena?

(a) Malic acid and acetic acid (b) Sulphuric acid and benzoic acid

(c) Citric acid and lactic acid (d) None of the above

24. When the pH reaches .... the milk protein, casein, coagulates as curd and that is known as iso-electric point of milk.

(a) 7.0 (b) 4.6

(c) 6.0 (d) None

25. Condensed milk is commonly used for...........

(a) Full cream unsweetened condensed milk.

(b) Full cream sweetened condensed milk.

(c) Unsweetened condensed skim milk.

(d) None of the above

26. Indian standard require that the moisture content of the dried milk be less than............ %.

(a) 3% (b) 4%

(c) 10% (d) 5%

27. SMP stands for ....

(a) Solid milk powder (c) Skim milk powder

(c) Sweetened milk products (d) Standard milking practice.

28. When milk is coagulated with rennet or pepsin according to conventional practise, it creates a soft curd.

(a) Soft curd milk (b) Yoghurt

(c) Homogenized milk (d) Humanized milk

29. The milk which has been heated for a temperature above 100°C for such lengths of time that it remains fit for consumption for at least 7 days at room temperature is known as

(a) Sterilized milk (b) Soft curd milk

(c) Pasteurized milk (d) None of the above

30. When we combine the liquor of a mash, made up ground barley and wheat flour with whole milk and then dry the moisture, we get

(a) Sweetened and condensed milk (b) Infant milk

(c) Malted milk (d) Fortified milk.

31. SMP has lower energy value, higher protein, calcium and riboflavin content as compared to dried whole milk and is devoid of.

(a) Vitamin A (b) Vitamin C

(c) Vitamin B (d) None

32. **According to PFA rules, khoa is the product obtained from cow or buffalo (or goat) milk or a combination thereof by rapid drying and the milk fat is .... the finished product.**

(a) Not less than 40% (b) Not less than 20 %

(c) Not more than 20% (d) None of the above

33. **The toned milk has at least**

(a) 3% fat and 8.5% SNF (b) 3% fat and 9% SNF

(c) 1.5% fat and 8.5% SNF (d) 1.5% fat and 9% SNF

34. **Indian standards require that the sweetened condensed milk should have .... per cent total milk solids and per cent fat.**

(a) 31, 9 (b) 41, 23

(c) 56, 12 (d) 25, 25

35. **When cow or buffalo milk is so modified in its chemical composition that it resembles human milk, then it is called**

(a) Modified milk (b) Humanized milk

(c) Imitation milk (d) All of these

36. **Malai should not contain ....milk fat, according to PFA rules.**

(a) Less than 15 % (b) More than 25%

(c) Less than 25% (d) More than 66%

37. **The minimum fat percent in ghee is**

(a) 78% (b) 95%

(c) 90% (d) 99%

38. **Makkhan refers to the country / desi butter normally obtained by churning whole milk curd and should not have less than .... fat.**

(a) 40% (b) 60%

(c) 76% (d) 95%

39. **The drained water obtained as a by-product of cheese making is known as**

(a) Chhena (b) Whey

(c) Butter milk (d) Malai

40. **.................. vitamin is deficient in milk?**

(a) Vitamin A (b) Vitamin B

(c) Vitamin C (d) Vitamin K

41. The essential fatty acids present in milk fat, which gives the characteristic flavor are

(a) Linoleic and arachidonic acids (b) Linoleic and oleic acid.

(c) Lineleic and linolenic acid (d) All of above

42. The predominant sugar found in milk is...........

(a) Lactose (b) Sucrose

(c) Glucose (d) Hexose

43. ....has the highest lactose content and lower protein, fat and mineral contents than those in other milks.

(a) Cow's milk (b) Buffalo's milk

(c) Human milk (d) Camel milk

44. The amount of lactose in buffalo milk is about

(a) 3% (b) 5.2%

(c) 6.2% (d) 7.8%

45. The high ....content of buffalo milk is reflected in the higher calorific value.

(a) Protein (b) Fat

(c) Carbohydrate (d) Vitamin

46. The amount of fat in buffalo milk is about

(a) 4.6% (b) 5.6%

(c) 6.6% (d) 7.6%

47 he amount of protein in cow/buffalo milk is about

(a) 1-2% (b) 2-3%

(c) 3-4% (d) 4-5%

48. The water content in buffalo milk is about

(a) 82% (b) 84.2%

(c) 87% (d) 88.5%

49. The fat content in skim milk is

(a) 0.1% (b) 1.5%

(c) 0.5% (d) 2%

50. The fat content in butter milk is

(a) 82% (b) 0.4%

(c) 33% (d) None of above

51. The fat content in cheese is approx:

(a) 20% (b) 42%

(c) 30% (d) 62%

52. The fat content in butter oil is:

(a) 80% (b) 95-99%

(c) 90% (d) 99.5%

53. The energy supplied by milk fat is:

(a) 3 Cal/g (b) 9.3Cal/g

(c) 4.1 Cal/g (d) 11.4Cal/g

54. 100 g of cow milk normally gives .... energy.

(a) 75 Cal (b) 100 Cal

(c) 33 Cal (d) 300 Cal

55. The natural acidity of cow milk is usually ....

(a) 0.08-0.12% (b) 0.13-0.14%

(c) 0.16-0.18% (d) All of above

56. The pH value of buffalo milk is usually ....

(a) 5.6-5.8 (b) 6.1-6.3

(c) 6.7-6.8 (d) 7.2-7.3

57. The yellow colour of milk is due to the presence of

(a) Colloidal casein particles (b) Dispersed fat globules

(c) Carotene and xanthophylls (d) None of the above

58. .... is responsible for the flavor of milk?

(a) Lactose (b) Phospholipids and Minerals

(c) Fatty acids and fats (d) All of above

59. Skim milkcolouris

(a) Bluish yellow (b) Greenish yellow

(c) Creamy white (d) None of the above

60. There is stone formation in milk when

(a) The milk is cooled rapidly

(b) The milk is boiled rapidly

(c) The milk is evaporated at a low temperature

(d) None of the above

61. **The density and specific gravity of milk are usually given at**
(a) 15.6°C (b) 20°C
(c) 15.6°F (d) 20°F

62. **The specific gravity of cow milk is about**
(a) 0.93 (b) 1.03
(c) 1.3 (d) 1.78

63. **The specific gravity of milk fat is**
(a) 0.93 (b) 1.03
(c) 1.616 (d) 1.346

64. **The porosity of milk powder is**
(a) 20-32% (b) 43-51%
(c) 32-43% (d) 52-65%

65. **Lactometer is equipment used to determine purity of milk by knowing**
(a) Color of milk (b) Impurities in milk
(c) Specific gravity of milk (d) None of the above

66. **................can be used to determine the specific gravity of a solution?**
(a) Lactometer (b) Galactometer
(c) Hydrometer (d) All the above

67. **Doeffel's instrument is used for measurement of**
(a) Specific gravity of milk (b) Color of milk
(c) Physical impurities of milk (d) None of above

68. **The Gerber method is used for measurement of**
(a) Percentage of fat in milk
(b) Availability of lactose in milk
(c) Percentage total solids and SNF
(d) None of the above

69. **What is the amount of SNF in a milk sample having 3% fat and density of milk is 1016 kg/m3 20°C**
(a) 4.38% (b) 5.38%
(c) 8.38% (d) None of above

70. **The freezing point of fresh whole milk is approximately**
(a) 0.5°C (b) -2.5°C
(c) -1.5°C (d) -0.5°C

71. **The freezing point of cow milk is lower than that of water due to**
(a) Dispersion of fat (b) Dissolved solids
(c) Certain of minerals (d) None of the above

72. **As the concentration of solids in a solution increase**
(a) The freezing point reduces (b) The freezing point increases
(c) The boiling point reduces (d) None of the above

73 **The boiling point of normal milk at atmospheric pressure is**
(a) 100.05°C (b) 100.5°C
(c) 105°C (d) 100.15°C

74. **If the concentration of market milk is doubled, boiling point will approximately increase by**
(a) 0.5°C (b) 5°C
(c) 1.5°C (d) None of above

75. **The specific heat of whole milk is about**
(a) 0.93-0.94 kJ/kg-°K (b) 3.89-3.93 kJ/ kg-°K
(c) 0.93-0.94 kJ/kg-°K (d) 4.18 kJ/kg-°K

76. **The Lamb's equation is used for determination ....of milk.**
(a) Thermal conductivity (b) Specific heat
(c) Specific gravity (d) None of the above

77. **The viscosity of skim milk at 20-40°C temperature will be usually**
(a) 0.5-0.7 mPa.s (b) 1.03-1.7 mPa.s
(c) 2.9-3.5 mPa.s (d) 9.8-10 mPa.s

78. **..................process increases the viscosity of milk?**
(a) Homogenization (b) Ageing at low temperature
(c) Concentration (d) All the above

79. **The surface tension of water is approximately**
(a) 72 mN/m (b) 7200 mN/m
(c) 720 mN/m (d) All of above

80. **The average refractive index of cow milk at 40°C is**
(a) 34.611 (b) 0.3461
(c) 1.3461 (d) 3.461

81. **Phosphatase test is used for testing.**
(a) Testing suitability of milk for sterilization
(b) Homogenization efficiency
(c) Butter consistency
(d) Adequacy of pasteurization of milk

82. **Which of the following conveyors are normally used for conveying of milk cans?**
(a) Magnetic conveyors (b) Belt conveyors
(c) Roller conveyors (d) Chain conveyors

83. **CIP is a term associated with**
(a) Cleaning of equipment (b) Quality analysis of milk
(c) Standard of milk (d) All of these

84. **To inhibits the growth of bacteria in milk, the temperature of milk in the receiving section is maintained at about**
(a) 4°C (b) 0°C
(c) 12°C (d) 8°C

85. **The bottom slope of the milk silos are kept at**
(a) 2% (b) 6%
(c) 10% (d) 14%

86. **The target microorganism considered during pasteurization of milk is**
(a) Clostridium botulinum (b) Steroptococcus lactis
(c) Micobacterium tuberculosis (d) All the above

87. **A process similar to pasteurization, but uses lowertemperatures to kill bacteria in milk is**
(a) Blanching (b) Themization
(c) Upperisation (d) None of the above

88. **The pasteurization time and temperature have been standardized so that**
(a) It does not reduce the cream line
(b) It does not inactivate phosphates
(c) It does not affect colour
(d) None of the above

89. **The standard requirement for HTST milk pasteurization is**
(a) 63°C-30 min (b) 72°C-15 s
(c) 88°C-3 s (d) None of above

90. The HTST pasterurisation process is normally intended to give........ reduction of the number of viable organisms of milk.

(a) 99.9% (b) 99.99%

(c) 99.999% (d) 99.9999%

91. The Milk fat is in .... form and hence is readily utilised in the human body.

(a) Emulsified (b) Fermented

(c) Oxidized (d) Reduced

92. The milk, whose fat and solids not fat (SNF) have been adjusted to a predetermined level, is known as

(a) Toned milk (b) Standardized milk

(c) Recombined milk (d) Reconstituted milk

93. Product obtained when butter oil, skim milk powder and water are combined in the correct proportions is known as

(a) Standardized milk (b) Recombined milk

(c) Reconstituted milk (d) Toned milk

94. According to PFA rules, cream, excluding, sterilized cream is the product of cow or buffalo milk or combination thereof, which contains not less than

(a) 10% milk fat (b) 35% milk fat

(c) 25% milk fat (d) 85% milk fat

95. The optimum temperature for the formation of curd is ....

(a) 35-40°C (b) 15-20°C

(c) 25-30°C (d)55-60°C

96. The UHT pasteurization system for milk employs

(a) 110°C for 15-30 min (b) 88°C for 3 sec

(c) 88°C for 15-30 min (d) 110°C for 3 sec

97. Which of the following method of heat treatment gives the best texture and flavor of milk?

(a) Long hold vat type pasteurization

(b) HTST pasteurization

(c) UHT pasteurization

(d) UHT sterilization

98. **If the homogenizer is installed in the pasteurization line, the capacity is usually kept ....than the maximum flow rate of the system.**

(a) 1-2% higher (b) 10-12% higher
(c) 3-8% higher (d) 3-8% lower

99. **The UHT system of milk pasteurization usually have a ...type of heat exchanger.**

(a) Steam injection (b) Steam infusion
(c) Shell and tube (d) None of the above

100. **Uperization involves**

(a) Heating the milk to 118°C for 7-8 minutes
(b) Heating milk to 135°C for 10-15 sec
(c) Heating the milk to 150°C for less than 1 sec
(d) None of the above

101. **The off-flavor of milk can be removed by**

(a) Aeration
(b) Vacuum flashing or distillation
(c) Steam injection followed by vacuum washing
(d) All the above

102. **........... heat processing method is usually carried out at more than 100°C to kill almost all the microorganisms?**

(a) Pasteurization (b) Sterilization
(c) Blanching (d) None of the above

103. **The sterilization process involves**

(a) Killing of nearly all microorganisms
(b) Killing of selected microorganisms.
(c) Killing of all microorganisms
(d) Killing of all pathogenic microorganisms.

104. **The time and temperature requirement for in-bottlesterilization of milk is**

(a) 105-110°C for 10-15 min (b) 105-110°C for 30-45 min
(c) 135-150°C for 30-45 min (d) 135-150°C for 30-45 min

105. **The UHT or aseptic sterilization process of milk will involve**

(a) 135-150°C for 1-20 seconds (b) 135-150°C for 20-60 seconds
(c) 135-150°C for 1-2 minutes (d) 135-150°C for 3-5 minutes

106. The milk should be heated to.......before filtration/clarification.

(a) 33°C (b) 55°C

(c) 43°C (d) 60°C

107. In case of UHT sterilization,the pressure maintained in the chambers is usually

(a) 1-2 bar (b) 6-8 bar

(c) 12-16 bar (d) 100 bar

108. The objective of homogenization is to

(a) Reduce the size of fat globules

(b) To disperse fat globules uniformly in milk

(c) To have uniformity in size of fat globules

(d) All the above

109. .... is known as the heart of homogenizer?

(a) Pump (b) Breaker ring

(c) Homogenizing valve (d) Tension spring

110. .... metal is used for the homogenizing valve

(a) Nickel-chromium alloy (b) Stellite

(c) High carbon steel (d) All of above

111. Which of the following heat exchangers are most common for milk pasteurization

(a) Shell and tube type (b) Steam injection type

(c) Plate type (d) Steam infusion type

112. The flow rate in the heater and holder in the milk pasteurizer are regulated by a

(a) Timing pump (b) Restricter

(c) Pressure switch (d) All of the above

113. If there is pressure drop on the re-generator, then

(a) The pumps do not operate efficiently

(b) Air mixes with milk and causes burn on

(c) There is considerable de-stabilisation of product

(d) All of the above

114. The homogenization pressure is normally kept at

(a) 72-124 kg/cm^2 (b) 212-340 kg/cm^2

(c) 136-170 kg/cm^2 (d) all of these

115. A three-plunger positive displacement pump, or, triplex pump is used in homogenizer mainly to

(a) Increase the capacity of homogenizer

(b) Have overlapping strokes for uniform pressure.

(c) Improve the dispersion of fat globules

(d) Increase the life of the homogenizing valve.

116. The temperature of milk during the homogenization is kept at a minimum of

(a) 55°C (b) 65°C

(c) 33°C (d) None of the above

117. Reduced break up and more resistance to flow during homogenization may be due to……

(a) High temperature of milk during homogenization

(b) High temperature and pressure

(c) Low temperature of milk during homogenization

(d) Increased acidity

118. In commercial installations, the cream separation from milk is done usually by

(a) Sedimentation (b) Centrifugation

(c) Settling (d) None of the above

119. Which of the following is used as the raw material for obtaining cream?

(a) Whole milk (b) Skim milk

(c) Ghee (d) Butter

120. The centrifuges are used for

(a) Separation of immiscible liquids

(b) Removal of solids from a dispersion

(c) Clarification of liquids by removal of small amounts of solids

(d) All of the above

121. As per the standards, butter should contain at least ………… fat.

(a) 80% (b) 99%

(c) 66% (d) 90%

122. As per PFA rules, the butter should not contain more than …to common salt.

(a) 0.5 (b) 1.5

(c) 3 (d) 5

123. During the butter making process, one of the first steps is the pasteurization of cream at a temperature of

(a) 63°C (b) 82-88°C

(c) 72°C (d) 92-98°C

124. Most of the flavor development of butter take place when the pH is kept between

(a) 7-6 (b) 5.5-4.6

(c) 6-5 (d) 4.5-4.0

125. During ageing,................is added butter for flavor?

(a) Malic acid (b) acetic acid

(c) Citric acid (d) curd

126. During butter making, which of the following operation helps the fat globules to adhere to form large masses and coagulation of butter grains?

(a) Ripening (b) Churning

(c) Ageing (d) Working

127. The optimum churning temperature for butter preparation is ....

(a) 4-6°C (b) 20-25°C

(c) 9-11°C (d) 33-38°C

128. In a continuous floatation butter churn, after the butter milk and residual milk solids are separated, the sequence of operations on the butter grains are

(a) Mixing-salting-deaeration (b) Salting-deaeration-mixing

(c) Deaeration-salting-mixing (d) Mixing-deaeration-salting

129. The churning efficiency should be at least

(a) 80% (b) 90%

(c) 95% (d) 99%

130. During butter making, the extra yield of materials obtained in butter after churning is given as

(a) Churn overrun (b) Composition overrun

(c) Package fill control (d) None of the above

131. During spray drying of milk, the separation of dried powder from the air is done by

(a) Gravity settling (b) Bag filter

(c) Cyclone separator (d) All the above

132. The drying time in a spray dryer is usually

(a) 5-100 sec (b) 5 min

(c) 2-3 sec (d) 2-3 min

133. The droplet size of the feed liquid in a spray dryer isabout

(a) 2-3 μm (b) 50-100 μm

(c) 20-30 μm (d) 400-500 μm

134. Which type of atomizer gives the maximum residence time in a spray dryer?

(a) High pressure nozzle (b) Centrifugal atomizer

(c) Two fluid atomizer (d) Ultrasonic nozzle atomizer

135. The centrifugal atomizers used in commercial spray dryers are usually about .... Mm in diameter and rotate at .... rev/min.

(a) 20-30, 1000-3000 (b) 200-300, 1000-3000

(c) 20-30, 5000-10000 (d) 200-300, 5000-10000

136. In a spray dryer the high pressure nozzle produces liquids of a.......... micron size and the centrifugal atomizers produce............. micron size.

(a) 100-300, 50-60 (b) 50-60, 10-15

(c) 50-60, 100-3000 (d) 10-15, 50-60

137. The drying surface temperature in case of drum drying is kept at

(a) Less than 60°C (b) Between 70-80°C

(c) Between 90-100°C (d) More than 100°C.

138. The residence time in a drum dryer is

(a) 30-60 sec (b) 10-15 sec

(c) 1-2 min (d) 3-4 min

139. Which among the following has the maximum heat utilization efficiency?

(a) Spray dryer (b) Drum dryer

(c) Hot air cabinet dryer (d) Hot air tunnel dryer

140. Bulk density of spray dried milk will be ....that of drum dried milk.

(a) Lower

(b) Higher

(c) Depends on the temperature used in drum dryer

(d) Same

141. The typical heat consumption in conduction dryers is

(a) 1500-2000 kJ/kg of water evaporated

(b) 2800-3200 kJ/kg of water evaporated.

(c) 4000-10000 kJ/kg of water evaporated

(d) 15000-20000 kJ/kg of water evaporated

Answer Key

1.	(d)	31.	(a)	61.	(a)	91.	(a)	121.	(a)
2.	(d)	32.	(c)	62.	(b)	92.	(b)	122.	(c)
3.	(c)	33.	(a)	63.	(a)	93.	(b)	123.	(b)
4.	(b)	34.	(a)	64.	(b)	94.	(c)	124.	(b)
5.	(b)	35.	(b)	65.	(c)	95.	(a)	125.	(c)
6.	(d)	36.	(c)	66.	(d)	96.	(b)	126.	(b)
7.	(b)	37.	(d)	67.	(a)	97.	(c)	127.	(c)
8.	(a)	38.	(c)	68.	(c)	98.	(c)	128.	(b)
9.	(a)	39.	(b)	69.	(b)	99.	(a)	129.	(d)
10.	(c)	40.	(c)	70.	(d)	100.	(c)	130.	(a)
11.	(a)	41.	(a)	71.	(b)	101.	(d)	131.	(d)
12.	(c)	42.	(a)	72.	(a)	102.	(b)	132.	(a)
13.	(a)	43.	(c)	73.	(b)	103.	(a)	133.	(b)
14.	(c)	44.	(b)	74.	(a)	104.	(b)	134.	(b)
15..	(b)	45.	(b)	75.	(b)	105.	(a)	135.	(d)
16.	(a)	46.	(c)	76.	(a)	106.	(c)	136.	(a)
17.	(b)	47.	(c)	77.	(b)	107.	(b)	137.	(d)
18.	(d)	48.	(b)	78.	(d)	108.	(d)	138.	(a)
19.	(b)	49.	(a)	79.	(a)	109.	(c)	139.	(b)
20.	(a)	50.	(b)	80.	(c)	110.	(b)	140.	(a)
21.	(c)	51.	(c)	81.	(d)	111.	(c)	141.	(b)
22.	(c)	52.	(d)	82.	(c)	112.	(d)		
23.	(c)	53.	(b)	83.	(a)	113.	(d)		
24.	(b)	54.	(a)	84.	(a)	114.	(c)		
25.	(b)	55.	(b)	85.	(b)	115.	(b)		
26.	(d)	56.	(c)	86.	(c)	116.	(b)		
27	(c)	57	(c)	87	(b)	117	(c)		
28	(a)	58	(d)	88	(a)	118	(b)		
29	(a)	59	(a)	89	(b)	119	(a)		
30	(c)	60	(b)	90	(c)	120	(d)		

Chapter-10

Food Processing Machinery

ONE LINER:

- Intermediate sieves with opening ratios of 1.139 or21/4would constitute a complete set.
- In gyratory screen, particle has circular motion at feed end and reciprocating motion at discharge end.
- Sorting of the material is based on the differences in size, shape, density, texture and color.
- In constant rate filtration the pressure drop is constant as filtration progress.
- The process in reverse osmosis operates at high pressure that of ultra-filtration.
- Expression is the removal of relatively small content of liquid from compressible milled food by mechanical means.
- Leaching refers to removal of soluble component from solid using a liquid solvent.
- Two important requirements a reverse osmosis membrane must meet for concentratingmilk and similar products are permeability, pH and temperature tolerance.
- Disc bowl centrifuge and Cyclone separatoris used to separate the materials by application of Centrifugal force.
- The shapes of opening in perforated metal screen are round, oblong or triangle

- The principle of operation of stone cleaner is based on specific gravity.
- The diameter of inlet is usually ¼th of diameter of cyclone.
- Filtration is a process of separating solid from liquids.
- In RO processing, energy required is approximately very low as compared to the energy required to evaporate water from milk.
- RO membrane pore size is between 0.1 to 1nm.
- Terminal velocity is the air velocity at which a particle remains in suspended state in a vertical pipe.
- Tyler sieves are based on 200 mesh screen.
- Pneumatic separation is based on the difference in density of various constituents of mixture.
- Grading of fruits and vegetables is necessary to group them as per their commercial value.
- Immiscible liquids are separated industrially in centrifugal decanters.
- Indented cylindersseparate grain on the basis of their relative length
- V' pocket type disc separator is used to separate round shaped material.
- Standard screens used to measure particle size ranges between76mm to 38 µm.
- Distillation utilizes the difference in vapor pressure to effect separation.
- IS 50 size screen has wider opening than IS 30 size screen.
- Disc separator separates the seed grains on the basis of grain length.
- A cyclone separator works on the principle of centrifugal and gravitational forces.
- Cream separator rotates at recommended speed of 6000 to 6500 rpm.
- A nine-fold increase in the radius of a centrifugal separator is equivalent to threefold increase in the angular speed of the separator.
- The gyratory screen is used with V-belt in horizontal screening surface.
- Gyrating and circular screens are rotary type screens.
- The specific resistance of a filter cake varies with the pressure drop.
- A cyclone separator works on the principle of centrifugal and gravitational forces.

- Stoke's law is used for determining the terminal velocity of particle.
- Micro-filtration involves separation of particles of 0.1 to 10 um size.
- The weight of hull is 18 to 22% of the total weight of paddy grain
- The endosperm weighs about 70 to 72%, whereas 4 to 6% bran is removed during polishing.
- The most commonly used solvent in the solvent extraction plant is n-hexane
- The de-oiled cake coming out of the extraction unit has 25 to 40% Solvent by weight.
- The optimum moisture content for paddy harvesting, storage and milling is22%, 12% and 14 %.
- Pulses contain 22 per cent of protein by weight.
- Usually raw paddy is dried from 20 to 22 percent moisture content to 14 to 16 per cent moisture content for higher head yield.
- Residual oil in de-oiled cake obtained after solvent extraction is about 1%, Ball mills run at 65 to 80% of their critical speed.
- The weight of husk is about 18 to 22 % of the total weight of paddy grain.
- The percentage of edible oil in raw rice bran is 10 to 20 %.
- Parboiling is the latest pre-milling treatment which improves the recovery of the head rice.
- In rubber roll paddy de-husker, the de-husking occurs due to compression & shear.
- Rittinger assumed that energy required in size reduction process is proportional to surface area change.
- A ball mill operating at high speeds with large balls is dominated by impact forces.
- Bond's law is used to calculate work index. In hammer mill, food is disintegrated by impact forces.
- The second stage in homogenization is used to re-disperse fat globules.
- Homogenized milk has globules less than 2 μ in diameter.
- The average calorific value of paddy husk is 3000 kcal/kg.
- For removing fuzz from the cotton seeds, delinters are used.
- The present day's rice mill which is probably known as modern rice mill uses the rubber roller sheller.

- Parboiling of paddy is a hydrothermal treatment.
- From the most of the wheat mills, the yield of flour can be obtained about 75%.
- The energy required by Kick's law to reduce size of particles from 50 to 25 mm is the same as that require from 16 mm to 8 mm.
- Cotton seed contains 15 to 20 % oil, Oil content is 45to 50 % in the shelled groundnut, soybean contains 40 % protein and 20 % oil content.
- During conditioning, pulses are finally dried to have 10 to 12 % moisture content.
- Head rice recovery in parboiled paddy is higher due to gelatinization of starch.
- The heart of the homogenizer is known as Valve.
- The normal size of fat globules in raw milk is 2to6 microns.
- After the first stage homogenization, the 2nd stage homogenization is done to re-disperse fatglobules at a pressure of 20 to 40 kg/cm2.
- Operation speed of ball mill for wet and dry grinding is kept 65% and 80% of the critical speed.
- Percent oil left in cake by traditional method, power driven ghani, solvent extraction and hydraulic press is 13 to 15%, 10.5 to 11.5 %, 1% and 7 to 8%.
- Operating speed of burr grinder varies from 650 to 750rpm.
- During parboiling, the steam requirement is 2000kg/ton of paddy parboiled.
- Capacity of saw gin varies from 4 to 5 kg/saw/hour.
- All the grains above 3/4 grain size are considered as head rice. Solvent extraction gives more than 99 per cent oil recovery.
- Hullers are traditionally used in rice milling and they seldom give above 75% total yield ofrice.
- The boiling point of n-hexane is 65.5°C and it is used as solvent in oil expression process.
- Moisture is added to the pulses at the rate of 3.5% for conditioning the pulses by alternate wetting and drying.

Questions

1. **Milling refers to**
   (a) Size reduction for converting raw commodity into edible form.
   (b) Size reduction and separation process for converting raw commodity into edible form.
   (c) Separation and cleaning process
   (d) None of the above

2. **..................equipment dehusks grains on the basis of compression and shear?**
   (a) Hollander (b) Centrifugal dehusker
   (c) Concave type husker (d) None of the above

3. **Which of the following equipment has a rapidly rotating abrasive roll encaged in a slowly rotating steel shell and is mostly used for milling of barley and wheat, etc.?**
   (a) Horizontal emery polisher (b) Hollander
   (c) Under runner disk sheller (d) None of the above

4. **Which of the following relationship is true**
   (a) ηhulling = Ehulling ×Ewk (b) Ehulling= ηhulling × Ewk
   (c) Ewk = ηhulling × Ehulling (d) None of above

5. **In a hammer mill, the hammers are rotated at .... rev/min.**
   (a) 200-300 (b) 600-700
   (c) 1500-4000 (d) 9000-10000

6. **Purees, food pastes, pulp and other similar products are processed by:**
   (a) Colloid mills
   (b) Ball mill
   (c) Concentric cylinder abrasive mill
   (D) Reitz mill

7. **Cryogenics can be advantageous for**
(a) Rice milling (b) Oil milling
(c) Spice milling (d) Pulses milling

8. **The roller mill works on the principle of**
(a) Shear and impact (b) Shear and compression
(c) Compression and friction (d) None of above

9. **The attrition mill works on the principle of**
(a) Shear and impact (b) Shear and compression
(c) Abrasion and compression (d) None of above

10. **The bowl mills and rod mills come under the' category of**
(a) Crushers (b) Intermediate and fine grinders
(c) Ultrafine grinder (d) Attrition mills

11. **Burr mill and plate mills are**
(a) Friction mills (b) Attrition mills
(c) Abrasive mills (d) None of the above

12. **The fluid energy mills come under the category or**
(a) Crushers (b) Intermediate and fine grinders
(c) Ultrafine grinder (d) Attrition mills

13. **To get very fine products from soft non-abrasive materials, size reduction mechanism is used?**
(a) Compression (b) Impact
(c) Attrition (d) Cutting

14. **The jaw crusher works mainly by the**
(a) Impact force (b) Shear force
(c) Friction force (d) Compression force

15. **The angle between two jaws in a jaw crusher is**
(a) 10-20° (b) 20-30°
(c) 30-40° (d) 60-70°

16. **The number of strokes in a jaw crusher is usually**
(a) 30-40/min (b) 100-200/min
(c) 250-400/min (d) 600-700/min

17. If in a pneumatic separator the fan is placed at the air exit, then the objective is to

(a) Create negative pressure within the chamber

(b) Create positive pressure in the chamber

(c) Create clean environment within the equipment

(d) None of the above

18. A cyclone separator is used for separating

(a) Solid particles from liquids (b) Liquid droplets from gases.

(c) Fine particles from gases (d) All of the above.

19. ....uses cyclone separator in a rice mill?

(a) Separation of bran conveyed by a bran aspirator

(b) Separation of husk from brown rice

(c) Separation of paddy from brown rice in a paddy separator

(d) None of the above

20. The gyratory crusher normally has ....

(a) 10-20 gyrations/sec (b) 10-20 gyrations/min

(c) 125-200 gyrations/sec (d) 125-200 gyrations/min

21. The blake crusher is a type of

(a) Jaw crusher (b) Gyratory crusher

(c) Smooth roll crusher (d) Toothed roll crusher

22. ....has a conical crushing head in a funnel shaped casing?

(a) Jaw crusher (b) Gyratory crusher

(c) Smooth roll crusher (d) Toothed roll crusher

23. ....refers to initial removal of large impurities in a processing operation.

(a) Scalping (b) Grading

(c) Sorting (d) Cleaning

24. The classification of commodities into various fractions depending on commercial value and usage is known as

(a) Scalping (b) Grading

(c) Sorting (d) Cleaning

25. Grading is used for

(a) Fruits and vegetables (b) Processed products

(c) Grains (d) All of these

26. Screening involves separation on the basis of

(a) Difference in size
(b) Difference in shape
(c) Difference in length
(d) None of the above

27. A screen denoted by 9V has holes

(a) In triangle form with 9 mm side
(b) In triangle form as 9 mm inscribed diameter
(d) All of the above
(d) In triangle form with 9 mm circumscribed diameter

28. Bond's size reduction law states that the work required to form particles of size "Dp' from very large feed is proportional to

(a) Mean particle size of the feed
(b) Square root of the surtace-to-volume ratio of the product.
(c) Square root of the volume to surface ratio of the product
(d) Square root of the final particle size of the product.

29. The work index is related to

(a) Kick's law
(b) Bond's law
(c) Rettinger's law
(d) None of the above

30. In size reduction of fine powders,............laws is most applicable?

(a) Kick's Law
(b) Rittinger's Law
(c) Bond's Law
(d) All of the above

31. In a size reduction machine, the sizes of the openings for the feed and product are 100 mm and 2 mm,respectively. The reduction ratio of the machine will be

(a) 50: 1
(b) 49: 1
(c) 2500:1
(d) Can't be calculated from the given data

32. The variable aperture screens can have

(a) Step wise increase in aperture from feed end
(b) Increase in aperture from feed end.
(c) Stepwise decrease in aperture from feed end
(d) Both (a) and (b)

33. A........... is a grid of parallel metal bars set on an inclined stationary frame.

(a) Grizzly (b) Trommel

(c) Plansifter (d) None of the above

34. The Dockins seed grader separates on the basis of difference in

(a) Length

(b) Color

(c) Seed thickness or cross sectional diameter

(d) Shape

35. Which of the following size reduction machine causes minimum heating of the food material?

(a) Gyratory crusher (b) Hammer mill

(c) Attrition mill (d) Ball mill

36. ....law states that within the elastic limit the work required for size reduction of a specific quantity of material is constant for the same reduction ratio, and it is independent of the original sizes?

(a) Kick (b) Bond

(c) Rettinger (d) All the above laws

37. Which of the following law states that the work is proportional to the square root of the surface area to volume ratio of the product?

(a) Kick's (b) Rettinger's

(c) Bond's (d) All the above

38. The Kick's law is given as E = In (L1/L2). In this equation, E has a unit of

(a) kW (b) kWh

(c) kWh/tonne (d) E is dimensionless and is just a number

39. Grizzly is a

(a) Fixed aperture screen

(b) Variable aperture screen.

(c) Can be either fixed aperture or variable aperture type

(d) None of the above

40. In case of a grizzly

(a) The bars are usually across the slope

(b) The bars are along the slope.

(c) The bars can be either along or across the slope

(d) None of the above

41. The disc separator separates on the basis of differences in

(a) Length (b) Width

(c) Hickness (d) Specific gravity

42. The triure is the other name of

(a) Disc separator (b) Velvet roll separator

(c) Indented cylinder separator (d) None of the above

43. The V-pockets in a disc separator is used for separating

(a) Flat seeds (b) Rice

(c) Tubular and round seeds (d) None of the above

44. For separating rice grains, the disc separator uses .... type pockets.

(a) A (b) R

(c) E (d) V5

45. Rettinger assumed that the work required in crushing is proportional to

(a) The linear dimension of the product

(b) The surface area to volume ratio of product

(c) Square of the common linear dimension

(d) None of the above

46. Which of the following law assumes that the size reduction is essentially a shearing procedure?

(a) Kick's (b) Rettinger's

(c) Bond's (d) all the above

47. The ratio of surface energy created by crushing to the energy absorbed by the solid is known,

(a) Work index (b) Energy efficiency ratio

(c) Crushing efficiency (d) Size reduction factor

48. The Bond's law is related to

(a) Separation effectiveness (b) Size reduction
(c) Milling of wheat (d) Oil expression

49. ....law is based on stress analysis of plastic deformation within the elastic limit?

(a) Kick's (b) Rettinger's
(c) Bond's (d) all the above

50. If g is the gap between the rolls and R is the roll radius, then the maximum size of particle (dp) in smooth roll crusher is given as

(a) dp= 0.04R + (g/2) (b) dp = 0.06R +(g/2)
(c) dp= 0.04R+ g (d) dp=0.06R +g

51. In a roll mill, the diameter of rolls are 25 cm and the between rolls is 10 mm. Then the maximum thickness of the grain that can enter through the rolls

(a) 12 mm (b) 15 mm
(c) 18 mm (d) 20 mm

52. The toothed roll crushers, in which the corrugated rolls are rotated at different speeds are known as

(a) Burr mills (b) Hollanders
(c) Disintegrators (d) None of the above

53. Size reduction by serrated roll crushers are by

(a) Compression and shear
(b) Impact, compression and shear
(c) Shear and impact
(d) Only compression

54. The hammer mill works mostly on the principle of

(a) Shear (b) Compression
(c) Impact (d) None of the above

55. The separation of husk reduces .... of the meal.

(a) Carbohydrate (b) Starch
(c) Fibre (d) Protein

56. The formula to determine the capacity of a pair of rolls has been given by

(a) Hollander (b) Kuprit
(c) Jacob (d) None of the above

57. **Normally in the Tyler sieves (adopted by US Bureau of Standards), the next sieve opening is............ times the previous screen size.**

(a) 1.12 (b) 1.25

(c) 1.44 (d) 1.5

58. **The fineness of grinding is controlled by............ In a hammer mill.**

(a) The rev/min of the rotating disc

(b) The type of blades.

(c) Size of openings on the screen

(d) All of these

59. **The ball mill works on the principle of**

(a) Compression (b) Impact

(c) Shear (d) Compression and sheaar

60. **Ball mills are generally used for**

(a) Coarse grinding of particles

(b) Ultra fine grinding of particles

(c) Medium to fine grinding of particles

(d) Coarse to ultra fine grinding of particles.

61. **Colloid mills are used for**

(a) Coarse grinding (b) Fine grinding

(c) Shearing (d) All of the above

62. **Normally in the Tyler sieves (adopted by US Bureau of Standards), the opening size of sieves is based on**

(a) 100 (b) 150

(c) 200 (d) None of the above

63. **The work index is related to**

(a) Size reduction (b) Oil milling

(c) Filtration (d) Wheat milling

64. **If a wire mesh is denoted as 8 mesh then the number of openings in a square inch is**

(a) 8 (b) 32

(c) 64 (d) None of these

65. ....is the minimum clear space between the edges of the opening on a screen surface.

(a) Mesh (b) Slot

(c) Aperture (d) None of the above

66. Air screen cleaners use ......for cleaning and for separation of different fractions of a mixture.

(a) Air pressure (b) Screening and aspiration

(c) Size separation on a screen (d) None of the above

67. Which of the following is used to improve the screen effectiveness?

(a) Scalper apron (b) Screen dams and screen knockers

(c) Oil cloth drape (d) All the above

68. Trommel is the other name of:

(a) Indented cylinder separator (b) Disk separator

(c) Revolving screen or cylinder (d) None ofabove

69. The oversize particles which do not pass through the Screen openings are called

(a) Undersize (b) Oversize

(c) Tails (d) None of the above

70. The aperture of a screen is

(a) The clear opening on a screen

(b) The distance between the ends of the screen.

(c) The diameter of holes in a screen

(d) None of the above.

71. Which of the following indicates the uniformity of grind in a reduced material?

(a) Degree of grinding (b) Work index

(c) Fineness modulus (d) None of the above

72. The fineness modulus (F) and average particle size D (in mm) are related as

(a) D=0.135(1.366)F (b) F= 1.366(0.135)D

(c) F= 0.135(1.366)D (d) D= 1.366(0.135)F

73. Destoner is used for separating stones from seed on the basis of

(a) Difference in terminal velocity (b) Difference in specific gravity

(c) Difference in flowability (d) None of the above

74. The floatation sorters work on the principle of differences in

(a) Size (b) Surface texture

(c) Specific gravity (d) None of the above

75. To separate mustard seed from wheat, the best type of separator is

(a) Indented cylinder separator (b) Specific gravity separator

(c) Centrifugal separator (d) Spiral separator

76. The major forces acting on a cyclone separator are

(a) Impact and frictional (b) Shear and impact

(c) Gravity and centrifugal (d) Frictional and centrifugal

77. In general, the height of the conical portion of a cyclone separator is ....times the height of the cylindrical section.

(a) 1-1.5 (b) 2-3

(c) 4-5 (d) None of the above

78. The electrostatic separator works on the principle of difference in

(a) Electrical resistance (b) Surface charge

(c) Electrical conductivity (d) None of the above

79. In a colour sorter, the defective pieces are separated usually by a

(a) Blast of compressed air (b) Electrical resistance mechanism

(c) Light mechanism (d) None of above

80. The magnetic separation can separate different fractions of a mixture on the basis of difference in

(a) Shape (b) Surface texture and stickiness

(c) Roundness (d) None of the above

81. The spiral separator separates on the basis of differences in

(a) Color (b) Size

(c) Surface texture and shape (d) None of the above

82. The velvet roll separator separates on the basis of differences in

(a) Shape and surface texture (b) Color and shape

(c) Shape and size (d) None of the above

83. Belt and roller sorter is used to sort the commodities as per differences in

(a) Size (b) Shape

(c) Specific gravity (d) None of the above

84. Which of the following uses the principle of separation depending on difference in aerodynamic properties?

(a) Paddy separator (b) Husk aspirator

(c) Polisher (d) None of the above

85. .....type separator is used for separation of milk powder from air coming out of a spray dryer.

(a) Cyclone (b) Filter cloth

(c) Inclined draper (d) None of the above

Answer Key

1.	(b)	21.	(a)	41.	(a)	61.	(b)	81.	(c)
2.	(c)	22.	(b)	42.	(c)	62	(c)	82.	(a)
3.	(b)	23.	(a)	43.	(c)	63.	(a)	83	(a)
4	(a)	24.	(b)	44.	(b)	64.	(c)	84.	(b)
5.	(c)	25.	(d)	45.	(c)	65.	(c)	85.	(a)
6.	(d)	26.	(a)	46.	(b)	66.	(b)		
7.	(c)	27.	(b)	47.	(c)	67.	(d)		
8.	(b)	28.	(b)	48.	(b)	68.	(c)		
9.	(c)	29.	(b)	49.	(a)	69	(c)		
10.	(b)	30.	(b)	50.	(a)	70.	(a)		
11.	(b)	31.	(a)	51.	(b)	71.	(c)		
12.	(c)	32.	(d)	52.	(c)	72.	(a)		
13.	(c)	33.	(a)	53.	(b)	73.	(b)		
14.	(d)	34.	(c)	54.	(c)	74.	(c)		
15.	(b)	35.	(b)	55.	(c)	75.	(d)		
16.	(c)	36.	(a)	56.	(b)	76.	(c)		
17.	(a)	37	(c)	57.	(c)	77.	(b)		
18.	(c)	38.	(c)	58.	(c)	78.	(c)		
19.	(a)	39.	(b)	59.	(b)	79.	(a)		
20.	(d)	40.	(c)	60.	(c)	80.	(b)		

Chapter-11

Agricultural Structures

ONE LINER:

- The indentation marks left on bricks during the process of moulding are known frog.
- The proportion of cement mortar used for pointing work and plastering work is 1:2 and 1:3.
- The compressive strength of a standard brick should be 35 kg/cm^2.
- Average size family of five members, the septic tank capacity may be kept about 2.8 m^3. For every additional member, the capacity should be increased by 0.42 m^3.
- The capacity of the tank should not be smaller than 2.1 m^3.
- Barbed wire is made of 14 gauge wire, available in a roll of 50 kg stretching to about 160 m long.
- The voltage applied to the electric fence varies between 6000 and 15000 volts under different conditions.
- In an electric fencing, from the point of view of safety, the circuit may be kept on for about 1/10thof a second and off for about 4/5thof a second.
- The maximum value of the current in electric fencing must not exceed 0.008 ampere, and a pulsating current of 0.005 ampere is considered satisfactory.
- The maximum length of the barn should be restricted to 40m so that nearly 50 cows can be housed in one barn unit.

- Gutters in the stanchion barn are usually 45 cm wide and 15 cm deep.
- Scum is a partially submerged mat of floating solids formed at the surface in a septic tank.
- Body weight of a Murrah buffalo on the average is 680 kg.
- Cage houses are built in warm regions where birds need no protection from the cold winds.
- Pit silos are made in areas where the soil is deep and the water table is very low.
- Generally, single storied poultry houses are preferred in warm regions.
- A floor area of 0.36 sq m per bird is usually provided and floor area of 0.38-sq. meter per bird is usually provided in deep litter system.
- Connecting pipe for septic tank should not be less than 100mm.
- The capacity of septic tank for 100 users will be roughly7-8 cum.
- The water consumption per head for domestic purposes for average conditions is taken as 135 liter/day.
- The permissible pH value and hardness limit of potable water is 7-8.5 and 75-115ppm.
- The total count of bacteria in potable water should not exceed 100 per cc.
- In case of safe drinking water, the Biochemical Oxygen Demand (BOD) should be nil.
- The minimum diameter of sewer pipe and manhole opening should be 15 cm and 50 cm.
- Specific gravity of most of the stones is usually in the range 2.4-2.8.
- In a good quality stone, the moisture absorption should be limited to 5%.
- The weight density of common bricks varies from 1600-1920 kg/cm3while crushing strength of a first class brick is 105 kg/cm^2.
- Water absorption by a first class bricks in 24 hours immersion in cold water should not exceed 20 %.
- The moisture content of a well seasoned wood is in the range 10-12%.
- Loose housing barn is termed as loafing barn system.
- Septic tank should be cleaned once in 2 years and cleaned before the sludge level reaches the bottom of outlet.
- The back of the manger is about 60 cm above the floor so that cow's may not throw the fodder into the feed alley.

- For adequate lighting in barn, 0.37 m^2 window area per cow is desirable.
- Milk production declines one standard deviation or more between a THI of 76 and 77.
- Environmentally controlled summer shelters for dairy cows should ideally provide a THI of less then 75.
- When Temperature Humidity Index (THI) reaches 70, 10% people feel uncomfortable & when THI reaches 79, 100% people feel uncomfortable.
- For designing manure pit, space requirement is calculated at 0.042 cum. per cow per day.
- Specific total heat production for broilers decreases as live weight increase.
- Average floor area required per animal for cow pen is 11 sq m and for buffalo pen is 14 sq m.
- 1 cubic meter of silage weighs 650 and cow is fed 3 kg of silage per 100 kg of body weight/day.
- In electric fencing, wooden/concrete posts are placed about 10 mapart and it is low cost fencing.
- Generally cow is fed with 14 to 18 kg of silage ration per day.
- Quantities for wood work are computed generally in terms of volume.
- The standard width of asbestos cement corrugated sheets is 1.05 m
- Minimum side lap in case of asbestos cement corrugated sheets is 4 m.
- End lap in case of asbestos cement corrugated sheets is 15 cm.
- The number of corrugations per sheet in case of asbestos cement corrugated sheets is 7.
- Exposure factor can be calculated using the relationship EF= (AtUav)/N
- Loose housing barn has 9-11 m2 per cow.
- Heat produced by dairy animals varies with body size and the environmental conditions.
- A barn in which cows run freely is known as loose house.
- In a stanchion barn, gutter is 45 cm wide, 15 cm deep with minimum slope of 2 %.
- A floor area of 0.36 square meters per bird is provided in deep litter poultry house.

- The spoilage of silage and dry matter loss in horizontal silos ranges between 20-30%
- Shrinkage cracks can be prevented in reinforced concrete water tank walls by slow drying out of concrete.
- Cage houses for poultry birds are generally built in warm regions where birds need no protection from the cold winds.
- As the number of birds or animals in a building increase, exposure factor decrease.
- For poultry housing in warm climates, window area should be 25 % of floor area.
- For 100 hen poultry for large breed, floor space requirement would be 37 m^2.
- Dimensions for single bird cage are approximately 0.6×0.2×0.45 m.
- Width of mangers is generally 35-90 cm.
- If sheep are housed together, nearly 1.12 m^2 of floor area is needed per sheep.
- In loose housing barn, floor space for loafing and feeding required is around 9-11 m^2 per cow.
- The total length of 6 lines of barbed wire for fencing 4 hectares square farm would be 4800 m.
- Total windows area in a poultry house in India should be about 25% offloor area.
- Mangers in a pen barn should be about 75 cm inside, with the top approximately 75 above the floor.
- The capacity of each pen in the poultry house should not exceed 250 birds.
- Mangers used in dairy barn should have widths ranging from 70-90 cm.
- In stanchion barn a slope of 2% is provided in gutters to drain the water.
- Types of wire-fencing for the farm are plain, barbed and electric.
- The brooding house should provide temperature of 32.2 °C to 37.7°C for three days old chicks.

Questions

1. **If gable roof is used in dairy barn, the maximum height of ceiling may vary between**
   (a) 5 m to 5.5 m (b) 5.5 m to 6.0 m
   (c) 4.5 m to 5.0 m (d) None of the above
2. **The back of the manger is about...... above the floor**
   (a) 40 cm (b) 50 cm
   (c) 60 cm (d) 80 cm
3. **Labour requirement is more in:**
   (a) Loose housing barn (b) Stanchion barn
   (c) Open air barn (d) All of the above
4. **The loose housing barn is one in which the cows are housed in a covered yard and**
   (a) They are also milked there
   (b) They are also milked in batches in the milking parlor
   (c) They are milked in separate area
   (d) They are not milked there
5. **Which type of housing is cheaper?**
   (a) Loose housing barn (b) Stanchion barn
   (c) Both has the same cost (d) None of the above
6. **Fire bricks are used:**
   (a) To reflect heat (b) To protect building against fire
   (c) To increase heat flow (d) To decrease heat flow
7. **Slump test for concrete is carried out to determine**
   (a) Durability (b) Strength
   (c) Water content (d) Workability

8. **The first class brick properties are**
   (a) Clean and even surface but not smooth
   (b) Dull sound
   (c) Crushing strength in between 35 to 75 kg/cm2
   (d) A little large voids

9. **A 3-ply plywood may have thickness of**
   (a) 12-15 mm (b) 9-12 mm
   (c) 6-10 mm (d) 3-6 mm

10. **For the RCC construction of water retaining structures, the grade concrete mix u generally:**
   (a) 1:2:4 (b) 1:1:2
   (c) 1: 1.5: 3 (d) 1:2:2

11. **UV stabilized polyethylene, on an average, transmits about .... of photo- synthetically active radiation (PAR) into the green house:**
   (a) 80% (b) 85%
   (c) 75% (d) 87%

12. **The basis for the design of water distribution system for a place is:**
   (a) Maximum weekly consumption
   (b) Average annual consumption
   (c) Peak hourly consumption plus the fire demand
   (d) Peak hourly consumption

13. **A water supply scheme is usually designed for**
   (a) 25 years (b) 50 years
   (c) 75 years (d) 100 years

14. **Which source gives surface water?**
   (a) Tube wells (b) Lakes
   (c) Rain (d) Springs

15. **In which type of barn, the cows are housed in a covered or partially covered yard, but they are milked in batches in a milking parlor**
   (a) Loose housing barn (b) Stanchion barn
   (c) Open air barn (d) All of these

16. **In a stall barn, the floor area (in m) required for each cow is**
   (a) 3.5 to 5.5 (b) 5.5 to 7.5
   (c) 7.6 to 9.5 (d) 9.6 to 11.6

17. **The air exchange caused by wind forces is given by**
(a) Q =1000 EAV
(b) Q= 1000 (EA/V)
(c) Q = 1000 (EV/A)
(d) Q= 1000 (AV/E)

18. **The most recent addition of green house film plastic covering is**
(a) Thermo-plastic
(b) Polyester
(c) Tefzel film
(d) Polyvinyl chloride

19. **A cooling tower cools waters up to temperature which as compared to wet bulb temperature is**
(a) Less
(b) Slightly more
(c) Equal
(d) None of the above

20. **Overall heat transfer coefficient unit's**
(a) W/m^2K
(b) W/m K
(c) Wm/K
(d) Wm^2/K

21. **CBR stands for:**
(a) Canadian Bearing Ratio
(b) Cutting Between Roads
(c) Central Bearing Ratio
(d) California Bearing Ratio

22. **The process used for winter air conditioning is**
(a) Heating & humidification
(b) Cooling & dehumidification
(c) Cooling & humidification
(d) Heating & dehumidification

23. **The air change required during hot weather to remove metabolic heat produced in livestock and poultry buildings**
(a) One air change per minute or more
(b) One air change per hour or more
(c) Two air changes per hour or more
(d) Four air change per hour or more

24. **Environmentally controlled summer shelters for dairy cows should ideally provide a THI of less than**
(a) 25
(b) 50
(c) 75
(d) 100

25. **Milk production declines one standard deviation or more when Thermal Humidity Index (THI) is between**
(a) 76 and 77
(b) 70 and 71
(c) 66 and 67
(d) 60 and 61

26. **Environmentally controlled summer shelters for dairy cows should ideally provide a Thermal Humidity Index (THI) of**
    (a) 100 (b) More than 75
    (c) Less than 75 (d) None of the above

27. **Even air distribution is achieved inside a building with**
    (a) Pressure ventilation (b) Exhaust ventilation
    (c) Natural ventilation (d) Flue ventilation

28. **A water closet is usually made of**
    (a) Cement concrete (b) Porcelain
    (c) Glazed earth ware (d) Stoneware

29. **A good building stone should have**
    (a) Strength (b) Good appearance and color
    (c) Resistance to fire (d) All of the above

30. **In brick masonry, the frog of the brick is normally kept on the**
    (a) Top face (b) Bottom face
    (c) Side face (d) All of the above

31. **Hollow bricks are used for:**
    (a) Earthquake proof buildings (b) Reducing the cost
    (c) Resistance against heat flow (d) Ornamental design

32. **Convective heat transfer follows**
    (a) Newton law (b) Fourier law
    (c) Stefan-Boltzmann law (d) Fick law

33. **During dehumidification of air, dry bulb temperature**
    (a) Increases (b) Decreases
    (c) Remain constant (d) Can increase or decrease

34. **Low temperature during concrete lying**
    (a) Decreases strength (b) Increases strength
    (c) Has no effect on strength (d) None of above

35. **Quantities for iron work are computed in terms of**
    (a) Number (b) Area in square meter
    (c) Volume in cubic meter (d) Weight in kg

36. **Air conditioning includes**
    (a) Heating & cooling (b) Dehumidifying
    (c) Removal of air impurities (d) All the above

37. Air velocity can be measured using a

(a) Thermometer (b) Psychrometer

(c) Anemometer (d) Hygrometer

38. Curing of concrete are done by

(a) Ponding method (b) Spraying method

(c) Covering with wet sand (d) Covering with moist cloth

39. The life of best quality ultraviolet (UV) Stabilized film when used as greenhouse covering material is

(a) Two years (b) Three years

(c) Four years (d) Five years

40. Fan and pad system of evaporation cooling can cool the temperature of the incoming air by about: (ASRB)

(a) 60% of the difference between dbt and wbt

(b) 80% of the difference between dbt and wbt

(c) 70% of the difference between dbt and wbt

(d) 90% of the difference between dbt and wbt

41. Non sweating animals adjust to high ambient temperature by increased rate of: (ASRB)

(a) Drinking water (b) Respiration

(c) Feeding (d) All the above

42. Relative humidity in a dairy barn should be maintained at

(a) 50-55% (b) 60-65%

(c) 75-80% (d) 85-90%

43. Active summer cooling in green houses is achieved by

(a) Convective cooling (b) Adiabatic cooling

(c) Evaporative cooling (d) Iso-thermic cooling

44. The thickness of glass widely used as greenhouse covering material is

(a) 2 mm (b) 4 mm

(c) 6 mm (d) 8 mm

45. The thickness of most widely used glass as green house covering material is

(a) 2 mm (b) 3 mm

(c) 4 mm (d) 5 mm

46. ....is not the common size of steel bar?
   (a) 16 mm (b) 20 mm
   (c) 22 mm (d) 28 mm

47. ....is not the standard length of asbestos cement corrugated sheets?
   (a) 1.5 m (b) 1.75 m
   (c) 2.0 m (d) 2.5 m

48. The comfort condition in air conditioning is at
   (a) 10°C DBT and 50% RH (b) 20°C DBT and 60% RH
   (c) 30°C DBT and 80% RH (d) 40°C DBT and 90% RH

49. When water is added to cement
   (a) Chemical reaction starts (b) Heat is absorbed
   (c) Heat is generated (d) None of the above

50. Full strength of concrete is achieved after (days)
   (a) 7 (b) 14
   (c) 21 (d) 28

51. For slabs and beams the grade of concrete mix generally used is
   (a) 1:2:4 (b) 1:3:6
   (c) 1:4:8 (d) 1:5:10

52. During heating and humidification process
   (a) Dry bulb temperature increases (b) Humidity ratio increase
   (c) Enthalpy increases (d) All the above

53. Sludge from septic tank should be removed in
   (a) 1 or 2 years (b) 4 years
   (c) 3 years (d) 5 years

54. For average family size of 2 adults and 3 children, septic tank capacity should be
   (a) 2.5 m^3 (b) 2.6 m^3
   (c) 2.7 m^3 (d) 2.8 m^3

55. The seasoning of timber is done to
   (a) Make it water proof (b) Increase its strength
   (c) Heat the timber (d) Expel the moisture from timber

56. The process used for summer air conditioning is

(a) Heating & humidification (b) Cooling & dehumidification

(c) Cooling & humidification (d) Heating & dehumidification

57. When bitumen is graded as 75/15, the figure 75 represents

(a) Softening point in °C (b) Pour point in °C

(c) Fire point in °C (d) Flash point in °C

58. Each standard cement bag has a volume of

(a) 0.032 cu meter (b) 0.035 cu meter

(c) 0.036 cu meter (d) 0.038 cu mete

59. The traditional green house glazing material all over the world is

(a) Glass (b) Plastic

(c) Film (d) All the above

60. The volume of standard brick is

(a) 2.378×10^{-3} cu m (b) 2.178×10^{-3} cu m

(c) 2.778×10^{-3} cu m (d) 2.578×10^{-3} cu m

61. Disinfection of drinking water involves removal of

(a) Bacteria (b) Turbidity

(c) Odor (d) Color

62. The main disadvantage of hard water is

(a) Bad taste (b) Foul smell

(c) More turbidity (d) Increased soap consumption

63. The maximum length of stall barn to house 72 cows should be

(a) 33 m (b) 40 m

(c) 50 m (d) 100 m

64. In which type of barn, the cows are housed and milk in the same building

(a) Loose housing barn (b) Open air barn

(c) Stanchion barn (d) None of the above

65. The average size of a bull shed is

(a) 15 m x 10 m (b) 10 m x 20 m

(c) 10 m x 5 m (d) 5 m x 15 m

66. Mangers used in barns should have width ranging from

(a) 5-10 cm (b) 25-30 cm

(c) 40-50 cm (d) 75-90 cm

67. The type of barn in which the cows are housed and milk in the same building is known as

(a) Loose housing barn (b) Stanchion barn

(c) Open air barn (d) All the above

68. Potassium permagnate is used for

(a) De-chlorination (b) Improving color

(c) Removing organic matter (d) Reducing hardness

69. When THI reaches 79, the percent of people feel uncomfortable are

(a) 100% (b) 90%

(c) 80% (d) 70%

70. The human body in a cooled space constitutes cooling load of

(a) Sensible heat only (b) Latent heat only

(c) Sensible and latent heat (d) Vaporization heat

71. The average capacity (cubic meter) of septic tank should not be less than

(a) 2.5 (b) 2.1

(c) 2.03 (d) 2.01

72. Cow is fed about ....of silage per 100 kg of her body weight per day

(a) 1 kg (b) 2 kg

(c) 3 kg (d) 4 kg

73. A normal silage ration (kg per cow per day) is

(a) 4 to 8 (b) 8 to 12

(c) 12 to 14 (d) 14 to 18

74. The diameter of pit silo is limited to

(a) 3 m (b) 4m

(c) 5 m (d) 6m

75. To prevent spoilage, silage should be removed at the rate of

(a) 5 cm/day (b) 10 cm/day

(c) 15 cm/day (d) 20 cm/day

76. From the management point of view, the farm stead should be located

(a) Near the centre of the farm (b) On one side of the farm

(c) Near a well in the farm (d) None of the above

77. The minimum diameter of a manhole opening is

(a) 10 cm (b) 20 cm

(c) 30 cm (d) 50 cm

78. The thickness of barbed wire which is used for fencing is

(a) 10 gauge (b) 12 gauge

(c) 14 gauge (d) 16 gauge

79. Thickness of walls of reinforced concrete water tank should not be less than

(a) 10 cm (b) 15 cm

(c) 18 cm (d) 20 cm

80. Water that contains substances which are undesirable for drinking is called

(a) Contaminated water (b) Polluted water

(c) Dirty water (d) Potable water

81. Water free from impurities may be obtained from

(a) Well (b) River

(c) Spring along the hill slope (d) Lake

82. The overall dimensions of wire floored poultry house of 400 birds having a service alley or 1.8 m width in the middle of the house is

(a) 20x 5.5 m (b) 23 x 5.5 m

(c) 23 x 4.5 m (d) 25 x 5.5 m

83. The dimensions of cage house to house one bird may be

(a) 0.6 x 0.2 x 0.45 cm (b) 0.6 x 0,25 x 0.45 cm

(c) 0.8 x 0.3 x 0.4 cm (d) 0.8 x 0.4 x 0.45 cm

84. Average daily water consumption of 100 poultry birds will be

(a) 30 liters (b) 40 liters

(c) 50 liters (d) 60 liters

85. he average size of brooder house is

(a) 3 x 3 m (b) 3.75 x 3 m

(c) 4.25 x 4 m (d) 3.5 x 3.5 m

86. **The term ewe refers to**
   (a) Female sheep (b) Female goat
   (c) Male sheep (d) Male goat

87. **Generally a cow of 100 kg body weight is fed silage about**
   (a) 3 kg/day (b) 6 kg/day
   (c) 9 kg/day (d) 12 kg/day

88. **Cows should be fed silage per day at the rate with respect to body weight**
   (a) 2% (b) 3%
   (c) 4% (d) 6%

89. **Trench silo has:**
   (a) Circular cross-section (b) Trapezoidal cross-section
   (c) Rectangular cross-section (d) Triangular cross-section

90. **If silage is to be fed 150 days per year and rate of feeding is 10 cm/day, depth of siloshould**
   (a) 10m (b) 15m
   (c) 20m (d) 30m

91. **.......... is not a type of horizontal silo?**
   (a) Surface (b) Pit
   (c) All are types of horizontal silo (d) Trench

92. **Cage houses for poultry are generally built in**
   (a) Cold regions (b) Moderate region
   (c) Warm regions (d) All of the above

93. **.................type of fencing is cheaper?**
   (a) Woven wire
   (b) Plain wire
   (c) The cost of all the above are same
   (d) Barbed wire

94. **In deep liter system, how many birds can be accommodated in 2.4mx 2.4m space?**
   (a) 12 (b) 14
   (c) 16 (d) 18

95. The depth of litter in the deep litter poultry housing is within:

(a) 10-15 cm (b) 15-20 cm

(c) 20-25 cm (d) 25-30 cm

96. On an average a hen requires floor area of (sq. ft.)

(a) 2.0 (b) 2.5

(c) 3.0 (d) 3.5

97. In wire floored poultry houses, bird takes rest in horizontally placed known as

(a) Perches (b) Cages

(c) Nests (d) Roosts

98. Normal weight of an egg in grams is

(a) 50-60 (b) 60-70

(c) 70-80 (d) 80-90

99. The floor area provided per bird is usually,.............in a deep litter poultry house,

(a) 0.1 m^2 (b) 0.4 m^2

(c) 1.0 m^2 (d) 1.4 m^2

100. The floor area (m2) per head in a deep litter poultry house is

(a) 0.1 (b) 0.4

(c) 0.8 (d) 1.2

101. Cage type poultry houses are preferably used for

(a) Layers (b) Boilers

(c) New born chicks (d) All the above

102. Which type of fencing is commonly used for poultry housing?

(a) Plain wire (b) Barbed wire

(c) Hexagonal wire (d) Electrical

103. Deep litter poultry housing has a litter of depth

(a) 5 cm (b) 20 cm

(c) 30 cm (d) 40 cm

104. Floor space required per bird for small sized birds when 100 birds are kept in a poultry house is

(a) 0.27 (b) 0.32

(c) 0.37 (d) 0.42

105. Which type of housing offers an excellent opportunity for the birds to burrow in it wne the air temperature is high and thereby cool themselves

(a) Wire floored poultry housing (b) Cage house

(c) Deep litter poultry housing (d) All of above

106. Floor space required per bird for large sized birds when 100 birds are kept in a poultry house is

(a) 0.27 (b) 0.32

(c) 0.37 (d) 0.42

107. The percent saving of floor area of barn for "Face in type barn as compared to the 'Face out type stanchion barn is

(a) 10 (b) 20

(c) 25 (d) 30

108. Gutters in stanchion barn are usually 45 cm wide and ....deep

(a) 15cm (b) 20cm

(c) 25cm (d) 30cm

109. The width of feed alley in stanchion type dairy barn is

(a) 0.4m (b) 0.8m

(c) 1.0m (d) 1.2m

110. The floor area provided per bird is usually,............in a deep litter poultry house

(a) 100 sq. cm (b) 2250 sq. cm

(c) 3600 sq. cm (d) 5000 sq. cm

111. Floor area required per hundred adult hens

(a) 30-42 m^2 (b) 7-17 m^2

(c) 17-27 m^2 (d) 42-50 m^2

112. The dimensions in (cm x cm x cm) of a cage to house one bird may be

(a) 20 x 10x 50 (b) 30 x 10 x 50

(c) 60 x 20x 45 (d) 30 20x 45

113. The height of walls running along the dairy barn length may be about

(a) 1.8 m (b) 2.1 m

(c) 2.4 m (d) 3.0 m

114. Common type of barn on modern dairy farms having large berds is

(a) Stanchion barn (b) Loose housing barn

(c) Open air barn (d) Milking parlor

115. The passage between outer wall and manger used to move push cart containing silage is called

(a) Feed alley (b) Trench

(c) Cart-way (d) Pen-alley

116. The gutters in the stanchion barn usually have dimensions of

(a) 40 cm x 15 cm (b) 42 cm x 12cm

(c) 45 cm x 15 cm (d) 45 cm x 12 cm

117. The requirement of floor space per cow in stall barn system of housing is

(a) 3.2-5.4 sq m (b) 5.5-7.5 sq m

(c) 7.2-8.4 sq m (d) 8.4-9.4 sq m

118. How much area (sq. ft.) is considered sufficient and economical for housing dairy animal comfortably?

(a) 20-25 (b) 25-30

(c) 30-35 (d) 35-40

119. The floor space required per cow both for loafing as well as for feeding in loose housing barn

(a) 9 to 11 sq m (b) 11 to 12 sq m

(c) 6 to 8 sq m (d) 12 to 14 sq m

120. The milk house should be separated from dairy barn by at least sage ....long paved passage

(a) 6m (b) 4m

(c) 8m (d) 10m

Answer Key

1.	(a)	31.	(c)	61.	(a)	91	(c)
2.	(c)	32.	(a)	62.	(d)	92.	(c)
3.	(b)	33.	(c)	63.	(b)	93.	(d)
4.	(b)	34.	(a)	64.	(c)	94.	(c)
5.	(a)	35.	(d)	65.	(a)	95.	(b)
6.	(d)	36.	(d)	66	(d)	96.	(d)
7.	(d)	37.	(c)	67.	(b)	97.	(d)
8.	(a)	38.	(d)	68.	(b)	98.	(c)
9.	(d)	39.	(c)	69.	(a)	99.	(b)
10.	(c)	40.	(b)	70.	(c)	100.	(b)
11.	(d)	41.	(b)	71.	(b)	101.	(a)
12.	(c)	42.	(c)	72.	(c)	102.	(c)
13.	(b)	43.	(c)	73.	(d)	103.	(b)
14.	(c)	44.	(b)	74.	(d)	104.	(b)
15.	(a)	45.	(c)	75.	(b)	105.	(c)
16.	(b)	46.	(d)	76.	(a)	106.	(c)
17.	(a)	47.	(d)	77.	(d)	107.	(b)
18.	(c)	48.	(b)	78.	(c)	108.	(a)
19.	(b)	49.	(c)	79.	(a)	109.	(d)
20.	(a)	50.	(d)	80.	(b)	110.	(c)
21.	(d)	51.	(a)	81.	(c)	111.	(a)
22.	(a)	52.	(d)	82.	(b)	112.	(c)
23.	(a)	53.	(a)	83.	(a)	113.	(c)
24.	(c)	54.	(d)	84.	(a)	114.	(a)
25.	(a)	55.	(d)	85.	(b)	115.	(a)
26.	(c)	56.	(b)	86.	(a)	116.	(c)
27.	(b)	57.	(a)	87.	(a)	117.	(b)
28.	(c)	58.	(b)	88.	(b)	118.	(b)
29.	(d)	59.	(a)	89.	(b)	119.	(b)
30.	(a)	60.	(b)	90.	(b)	120.	(a)

Chapter-12

Agri-Project Planning and Management

ONE LINER:

- Break-Even Analysis is technique used by production management and management accounts.
- Break-Even Chart is graphical representation of costs at various levels of activity.
- Break-Even point, the point at which neither profit nor loss is made.
- Variable costs are those costs which vary directly with level of output.
- Break-Even point is the ratio of fixed costs to the selling price per units minus variable costs.
- Break-Even Analysis is used to determine how much sales volume business needs to start making a profit.
- Material index (MI) is the ratio of weight of local market material used in the finished. Product to the weight of the finished product.
- Material index (MI) >1, location should be nearer to the source of raw material.
- Material index (MI) <1, then location nearer to market is advised.
- Operating pressure of fire tube boiler is 1-24 bar.
- Operating pressure of fire water boiler is 100-140 bar.

- Steam boiler are rated in Btu/h, kW or Boiler HP.
- Conversion of Boiler units (1Btu/h= 0.293W, 1 Boiler HP= 9.8 kW).
- Plant layout is the optimum arrangement of different facilities including human resource, plant and machinery, material etc.
- The basic type of flow pattern used in designing the layout are I-flow, L-flow,U-flow,S-flow and U-flow.
- I-flowpattern used to separate receiving and shipping area and L-flow pattern used when straight line flow chart is to be accommodated while U-flow pattern used for combination of receiving and dispatch.
- S- Flow pattern used when the production line is long and zigzag on the production floor is required.
- O- Flow pattern used when it is desired to terminate to determine the flow near where it is originated.
- Product/Line layout is developed for product focused systems. In which, only one product, or one type of product, is produced in a given area.
- Process/Functional layout is developed for process focused systems. The processing units are organized by functions into departments on the assumptions that certain skill and facilities are available in each department.
- Cellular/Group layout is special type of functional layout in which the facilities are clubbed together into cells. It is suitable for systems designed to use the concepts, principles and approach of group technology.
- Fixed Position layout is suitable for single, large, high cost components or products.
- SLP indicates Systematic Layout Planning.
- Sales management is the term applied to the process of distributing goods.
- An important objective in process costing is the evaluation of wastage.
- Variable cost tends to very directly with the volume of output.
- Hygiene design requirements vary to some extend depending upon food being processed.
- A combination of flocculation and coagulation is used to agglomerate small particles during waste water treatment in food industry.
- Sodium hydroxide is a powerful detergent.

- Microbial growth requires definite pH.
- Unit of BOD is mg/litre.
- Milk stone is the portion of food left over equipment surface during processing.
- Sales forecasting is defined as estimation of type quality and quantity of future sales.
- Time study standardizes the time taken by employer to perform industrial operation.
- Time study is useful to determine standard time for various operations.
- The P-chart is used for control of faction defective.
- Preventive maintenance is intended to prevent break downs.
- Variable cost tends to vary directly with the volume of output.
- The C chart is a quality control chart showing number of defects in sub-groups of constant size.
- PFSs- Process flow diagrams; P&IDs- Process and instrument drawings.
- Total revenue is the amount of product sold multiplied by the unit price charge.
- CPM- Critical path method; PERT- Project Evaluation and review Technique.
- FIFO- First-In-First-Out; LIFO- Last-In-First-Out; SIRO- Service-In-Random-Order.
- Indian merchant shipping act-1923; Workmen compassion act- 1923.
- Trade union act- 1926; Payment of wedge act- 1936.
- Industrial employment act- 1946; Factories act- 1948.
- Minimum wedges act-1948; Employ state insurance act-1948.
- Mines act- 1952; Employee provident fund act- 1952.
- Maternity benefit act- 1961.

Questions

1. **Net work scheduling related to**
   (a) Programming (b) Management
   (c) Planning (d) None of above
2. **Critical path analysis (CPM) is**
   (a) Deterministic in nature (b) Event oriented
   (c) Probabilistic in nature (d) Dynamic in nature
3. **Decision support septum's are concerned with**
   (a) Unit operations (b) Marketing problems
   (c) Storage problems (d) Project planning
4. **When a septum of simultaneous linear equation is solved by using two phase simplex method, the value of decision variable may be**
   (a) Positive or Negative (b) Negative
   (c) Positive (d) All of above
5. **Food safety standard followed by**
   (a) ISO 22000 (b) ISO 9000
   (c) ISO 14000 (d) ISO 9001
6. **Food quality standard followed by**
   (a) ISO 22000 (b) ISO 9000
   (c) ISO 14000 (d) ISO 9001
7. **Codex Alimentarius commission established by FAO and WHO for setting of food standards in which year**
   (a) 1960 (b) 1963
   (c) 1980 (d) 1973
8. **Consumer protection act was passed in**
   (a) 1976 (b) 1956
   (c) 1986 (d) 1966

9. **Grading and marketing rules in respect of food established in year**
   (a) 1942 (b) 1911
   (c) 1937 (d) 1956
10. **Normal distribution curve shape is**
   (a) Bell shape (b) Parabolic
   (c) Hyperbolic (d) Rectangular
11. **Network scheduling is term related to**
   (a) Planning (b) Management
   (c) Scheduling (d) Programming
12. **"HACCP" stands for**
   (a) Hazardous analysis of critical control point
   (b) Hazard analysis of critical control point
   (c) Hazard analysis and critical control point
   (d) Hazard analysis and common control point
13. **Minimization of project duration can be done by**
   (a) Linear program (b) Network analysis
   (c) Optimum (d) Optimization
14. **Break even chart used for**
   (a) Cost volume profit
   (b) Impact of the rate of change in sales
   (c) All of about
   (d) None
15. **Dynamic programming deals with**
   (a) Multi-stage decision making problems
   (b) Time dependent decision making problems
   (c) Single stage decision making problems
   (d) Time independent decision making problems
16. **When a system of simultaneous linear equation is solved by using phase simplex method, the decision variable**
   (a) +ve or –ve (b) –ve
   (c) +ve (d) All of above

17. **The use of dummy activity network diagram when**
    (a) Two parallel act work have the same to one and head events
    (b) A & C
    (c) The chain of activities may have a common event
    (d) None of above

18. **ISO-14000 related to**
    (a) Environmental management (b) Population control
    (c) Food industries (d) Pollution control

19. **Dummy activity indicates**
    (a) Critical Path (b) Zero slack
    (c) Project planning (d) Active slack

20. **In critical path analysis, CPM is in nature**
    (a) Probabilistic (b) Static
    (c) Dynamic (d) Deterministic

21. **Codex Alimentarius commission, established in**
    (a) 1961 (b) 1960
    (c) 1963 (d) 1968

22. **Activity slack time used for**
    (a) Independent float (b) Total float
    (c) Free float (d) all of the above

23. **Environmental Food Protection act was formulated in**
    (a) 1986 (b) 1926
    (c) 1966 (d) 1937

24. **"ISO" standards behaves as like**
    (a) Non mandatory order (b) Non mandatory regulation
    (c) Mandatory orders (d) Mandatory regulations

25. **Linear programming problem must have**
    (a) Decision variables that one needs to determine
    (b) Constraints was one needs to specify
    (c) Objective that one aims to maximize or minimize
    (d) All of above

26. Network scheduling related to

(a) Management (b) Programming

(c) Planning (d) All of above

27. Dynamic programming deals the

(a) Time dependent decision making problems

(b) Time independent decision making problem

(c) Multi-stage decision making problems

(d) Single stage decision making problems

28. Break even chart used for

(a) Determining the impact of the rate of change in sales

(b) Determining the cost volume profit

(c) All of above

(d) None of these

29. Role of artificial variables in simple method is

(a) To start phase of simplex method

(b) To aid in finding initial basic feasible solution

(c) To find shadow price from final simplex method

(d) All of these

30. Network problem have advantages in

(a) Project scheduling (b) Project Planning

(c) Project Controlling (d) All of above

31. Objective of network analysis is to minimize the

(a) Total cost (b) Total project duration

(c) Production delays (d) None of these

32. CPM is (NET 2016)

(a) Control parabola method (b) Control point method

(c) Critical path method (d) Control pin method

33. ISO-13153:2012 frame work with

(a) Management for environmental pollution

(b) Energy saving single family residential and commercial buildings

(c) Food and safety

(d) All of above

34. **FASSI is governed by central ministry of**
   (a) Agriculture (b) Law and justice
   (c) Health and family welfare (d) Food processing industries

35. **CPM technique is related to**
   (a) Agri project planning and management
   (b) Transportation of processed material
   (c) All of above
   (d) None of these

36. **Main component of transportation model is**
   (a) Source (b) Destination
   (c) Sales (d) Source and destination

37. **ISO 9000 family of standard of**
   (a) Quality management system (b) Management system
   (c) Quality system (d) Production system

38. **AGMARK food law related to**
   (a) Quality of products (b) Regulates production
   (c) Purity of products (d) Quality and purity of products

39. **PERT develops a measure of**
   (a) Time (b) Mean and standard deviation both
   (c) Cost (d) Only mean

40. **FSSAI act introduced in**
   (a) 2015 (b) 2000
   (c) 2011 (d) 2006

41. **Draft of FSSAI rules were issued in**
   (a) 2006 (b) 2011
   (c) 2001 (d) 2000

42. **ISO stands for**
   (a) Indian Safety Organization (b) International Safety Organization
   (c) Indian Standards Organization (d) International Organization for Standardization

43. **Food safety standards related to**
   (a) ISO 22000 (b) ISO 9001
   (c) ISO 14000 (d) ISO 2000

44. AGMARK stands for

(a) Agricultural Produce (Grading and Marketing) act, 1937

(b) Agricultural Production

(c) Agricultural Performance

(d) All of above

45. AGMARK is administered by

(a) Ministry of Food Processing Industries

(b) Department of Health

(c) Department of Agriculture

(d) Directorate of Marketing and Inspection

46. PFA Stands for

(a) Prevention of Food Adulteration

(b) Preservation of Food Addition

(c) Production of Food Adulteration

(d) Preservation of Food Adulteration

47. PFA act 1955 enforced by

(a) Department of Agriculture

(b) Ministry of Commerce

(c) Department of Health

(d) Ministry of Processing Industries

48. MMPO (1992) stands for

(a) Milk and Milk Production Order

(b) Milk and Milk Products Order

(c) Milk and Meat Products Order

(d) All of these

49. MFPO (1973) stands for

(a) Milk and Milk Products Order

(b) Meat and Milk Products Order

(c) Meat Food Products Order

(d) Milk and Food Products Order

50. CAC stands for

(a) Codex Alimantarius Commission

(b) Critical Association method

(c) Central Association of Consumers

(d) Consortium of Applied Chemists

51. Quality assurance in relation to food manufacture ensure

(a) Raw material are good quantity

(b) Raw material are good quality

(c) Food meets specification and standards

(d) All of above

52. CAC established by FAO and WHO in year

(a) 1956 (b) 1963

(c) 1983 (d) 1973

53. FSSAI act 2006 is operationalized on

(a) 5th January 2006 (b) 5th August 2011

(c) 15th January 2016 (d) 21th January 2007

54. No. of acts repealed by FSSAI act, 2006,

(a) 2 (b) 10

(c) 8 (d) 11

55. For the Regulation of Import into India introduced by

(a) Plant Quarantine Order, 2003

(b) Product Quarantine Order, 2003

(c) Production Quarantine Order, 2003

(d) All of above

56. Plant Quarantine (Regulation of Import into India) Order, 2003 involved by

(a) Department of Agriculture & Cooperation

(b) Ministry of Health and Family resource

(c) Ministry of Food Processing Industries

(d) All of above

57. Weight and measurement act was introduced in

(a) 2006 (b) 1976

(c) 1966 (d) 1956

58. Weight and measurement act,1976 comes under

(a) Ministry of Food Processing Industries

(b) Department of Agriculture & Cooperation

(c) Department of Legal Metrology

(d) All of above

59. Bureau of Indian Standards (BIS) introduced in

(a) 2006 (b) 1986

(c) 1999 (d) 1952

60. Export (Quality Control and Inspection)Act, 1963 comes under the

(a) Export Inspection Council of India

(b) Department of legal Metrology

(c) Department of consumer affairs

(d) Directorate of marketing and inspection

61. ISO 9000:2005 Quality management systems deals with

(a) Fundamentals and vocabulary

(b) Guidelines for performance improvements

(c) Customer satisfaction

(d) Requirements

62. PFA stands for

(a) Presentation of Food Adulteration

(b) Prevention of Food Adulteration

(c) Preservation of Food additives

(d) Prevention of Food Activities

63. PFA act 1954 comes under

(a) Department of Food processing industries

(b) Department of agriculture

(c) Department of Finance

(d) Department of Health

64. ISO 9001:2008 Quality management systems deals with

(a) Fundamentals and vocabulary

(b) Customer satisfaction

(c) Requirements

(d) Guidelines for performance improvements

65. **ISO 9000 standards was released in**
   (a) 1987 (b) 1995
   (c) 2005 (d) 1994

66. **Role of BIS 9000 is**
   (a) Founder member of ISO
   (b) Represents India in ISO
   (c) National standards body of India
   (d) None of above

67. **ISO 9001:2009 quality management systems deals with**
   (a) Fundamentals and vocabulary
   (b) Guidelines for performance improvements
   (c) Customer satisfaction
   (d) Requirements

68. **FPO stands for**
   (a) Feed products order (b) Food products order
   (c) Fish products order (d) Fruits products order

69. **FPO,1955 comes under**
   (a) Department of health
   (b) Ministry of food processing industries
   (c) Department of agriculture
   (d) Department of Finance

70. **ISO 22000 is a**
   (a) Food safety standard (b) Feed safety standard
   (c) Food satisfy standards (d) All of above

71. **ISO 9000 represents**
   (a) Quality and safety (b) Safety system
   (c) Quality system (d) All of above

72. **Edible Oils Packaging (Regulation) order introduced in**
   (a) 2006 (b) 1957
   (c) 1947 (d) 1998

73. **Vegetable Oil Products (Control) order in**
   (a) 1947 (b) 2006
   (c) 1998 (d) 1955

74. According to FSSAI, for TSS and fruit pulp/juice content in unsweetened juice having minimum standards

(a) Natural: 5%
(b) Natural: 100 %
(c) Natural: 25%
(d) Natural: 15%

75. For fruit pulp/juice and TSS content in lime RTS having minimum standards, according to FSSAI.

(a) 5%,15 %
(b) 5 %,10 %
(c) 15%,50 %
(d) 12%,5 %

76. FSSAI recommends, minimum standards for fruit pulp/juice and TSS content in Syrup are

(a) 5 %, 10%
(b) 65%,25 %
(c) 15%,10 %
(d) 15%,15 %

77. Marine product export development authority introduced in,

(a) 1985
(b) 1952
(c) 1962
(d) 1972

78. Consumer protection act comes under

(a) Directorate of plant protection
(b) Dept. of Consumer affairs
(c) Ministry of law and justice
(d) All of above

79. Essential commodities act (related to food) comes in

(a) 1957
(b) 1998
(c) 1955
(d) 2006

80. APEDA act established in

(a) 1963
(b) 1945
(c) 1923
(d) 2006

81. PERT stands for

(a) Program Estimation and Review Technique
(b) Program Evaluation and reporting technique
(c) Program Evaluation and Review Technique
(d) All of above

82. CPM is the

(a) Activity oriented technique
(b) Event oriented technique
(c) Time oriented technique
(d) Target oriented technique

83. **VAM stands for**
   (a) Value added method (b) Vogel's approximation method
   (c) Value assessment method (d) All of above

84. **Key element is also called**
   (a) Pivot (b) Surplus
   (c) Slack (d) Artificial

85. **MODI stands for**
   (a) Model index solution (b) Modified distribution method
   (c) Mendel's distribution method (d) Modern distribution

86. **BCWS stands for**
   (a) Budgeted cost of work study (b) Budgeted cost of work schemes
   (c) Budgeted cost of work studied (d) Budgeted cost of work scheduled

87. **The main purpose of a project plan is to**
   (a) Guide project execution (b) Reduce risk
   (c) Meet standards expectations (d) All of above

88. **Interference is the difference of**
   (a) Total float (b) Free float
   (c) Total and free float (d) All of above

89. **Slack is also called**
   (a) Float (b) Event
   (c) Activity (d) Time

90. **BOD stands for**
   (a) Biological oxygen demand (b) Biology oxygen demand
   (c) Bio-bubble oxygen demand (d) All of above

91. **COD stands for**
   (a) Chemical oxygen demand (b) Chemical ozone demand
   (c) Calcium oxygen demand (d) Chlorine oxygen demand

92. **Symbol of circle represents**
   (a) Storage (b) Operation
   (c) Transportation (d) Inspection

93. **Symbol of rectangular represents**
   (a) Storage (b) Transportation
   (c) Inspection (d) Operation

94. **Symbol of triangle represents**
    (a) Transportation (b) Storage
    (c) Operation (d) Inspection

95. **Symbol of arrow represents**
    (a) Transportation (b) Storage
    (c) Operation (d) Inspection

96. **MRP in inventory control refers to**
    (a) Material required planning
    (b) Minimum requirement planning
    (c) Maximum requirement planning
    (d) Material requirement planning

97. **Zimpro process is used for,**
    (a) Waste water treatment (b) Thermal condition
    (c) Radiation conditions (d) Chemical conditions

98. **S-flow occurs in**
    (a) Vertical flow system (b) Horizontal flow system
    (c) Incline flow system (d) A&B both

99. **Type of inventory model**
    (a) ABC analysis (b) MRP
    (c) EOQ (d) All of above

100. **Prevention of Food Adulteration act enforced in year**
    (a) 1995 (b) 1955
    (c) 1896 (d) 1966

Answer Key

1.	(b)	21.	(c)	41.	(b)	61.	(a)	81.	(c)
2.	(a)	22.	(d)	42.	(d)	62.	(b)	82.	(a)
3.	(d)	23.	(a)	43.	(a)	63.	(d)	83.	(b)
4.	(a)	24.	(b)	44.	(a)	64.	(c)	84.	(a)
5.	(a)	25.	(d)	45.	(d)	65.	(a)	85.	(d)
6.	(d)	26.	(c)	46.	(a)	66.	(a)	86.	(d)
7.	(b)	27.	(c)	47.	(c)	67.	(b)	87.	(a)
8.	(c)	28.	(c)	48.	(b)	68.	(d)	88.	(c)
9.	(c)	29.	(b)	49.	(c)	69.	(b)	89.	(a)
10.	(a)	30.	(d)	50.	(a)	70.	(a)	90.	(a)
11.	(b)	31.	(a)	51.	(c)	71.	(c)	91.	(a)
12.	(c)	32.	(c)	52.	(b)	72.	(d)	92.	(b)
13.	(b)	33.	(b)	53.	(b)	73.	(a)	93.	(c)
14.	(c)	34.	(c)	54.	(c)	74	(b)	94.	(b)
15.	(a)	35.	(a)	55.	(a)	75.	(b)	95.	(a)
16.	(c)	36.	(d)	56.	(a)	76.	(a)	96.	(d)
17.	(b)	37.	(a)	57.	(b)	77.	(d)	97.	(a)
18.	(a)	38.	(d)	58.	(c)	78.	(b)	98.	(d)
19.	(b)	39.	(b)	59.	(b)	79.	(c)	99.	(a)
20.	(d)	40.	(d)	60.	(a)	80.	(a)	100.	(b)

Chapter- 13

Instrumentation and Process Control

ONE LINER:

- Turbine flow meter is suitable for flow totalization.
- Thermocouple is used to give an indication for temperature change.
- Photovoltaic cell is an active transducer.
- A strip chart recorder is an inverse transducer.
- A resistance potentiometer is a zero-order instrument.
- In wire-wound strain gauges, the change in resistance on the application of strain is mainly due to the change in length and diameter of the wire.
- Thermocouples are active transducers.
- Air cored inductive transducers are used at higher frequencies.
- The size of air-cored transducers as compared with their iron cored counterparts is bigger.
- Rotameter is a variable head flow meter.
- The tube of the rotameter is tapered downwards.
- Sliding vane flow meter is a positive displacement device.
- Hot wire anemometer is used for clean fluids.
- Hot wire anemometer is used for gas velocities measurement.

- ❐ A rotameter can be used in a horizontal pipeline with modifications.
- ❐ Solids are ananalyzed using a polarography.
- ❐ Bellows, diaphragm and bourdon tube are mechanical pressure sensing element.
- ❐ pH meter has 2 cell.
- ❐ Mercury thermometer can be used to measure the temperature up to 350 °C.
- ❐ Helium gas constant volume thermometer is suitable for the measurement of a temperature of less than 0°C.
- ❐ Static error judges the accuracy of an instrument.
- ❐ Gradual departure of the instrument output from the calibrated value is drift.
- ❐ The arc of C type bourdon tube is 250°.
- ❐ Measurement of sub-zero Celsius temperature in industry is done most commonly by resistance thermometers.
- ❐ Pressure of 0.0001 absolute psi can be measured by Mcloid gauge.
- ❐ Psychrometer determines the humidity of gasses.
- ❐ A mercury barometer measures the atmospheric pressure.
- ❐ Continuous measurement of moisture in paper is done by High resistance Wheatstone's bridge circuit.
- ❐ Platinum-platinum + rhodium thermocouples is the most suitable for measuring a temperature of about 1600°C in an oxidizing atmosphere.
- ❐ Radiation pyrometer measure the high temperature, around 2000°C by incandescent gas mantles.

Questions

1. **Principle used by the pitot tube ....**
   (a) Bernoulli's Theorem (b) Thomson Equation
   (c) Boyles Law (d) Archimedes principle
2. **Kelvin bridge is uses null galvanometer method for the measurement of**
   (a) Capacitance (b) Resistance
   (c) Inductance (d) Q factor
3. **Which device is used for the measurement of resistance**
   (a) Voltmeters (b) Ammeters
   (c) Wattmeter (d) All the above
4. **Which of the following isa differential pressure flow meter?**
   (a) Flow nozzle (b) Orifice meter
   (c) Venturimeter (d) All of the above
5. **Principle which is used for the the measurement of barometric pressure is also used for**
   (a) Air Density (b) Humidity
   (c) Altitude (d) Temperature
6. **The barometer developed to avoid use of liquid was called**
   (a) Aneroid Barometer (b) Android barometer
   (c) Bourdon Tube barometer (d) Spring Suspended barometer
7. **Piezoelectric load cell are used for measurement in-motion weigh bridges and the principle employed is**
   (a) Conversion of pressure into change of resistance
   (b) Conversion of pressure into generation of electrical signal
   (c) Conversion of pressure into change of inductance
   (d) Conversion of pressure into change of capacitance

8. **Bourdon tube sensors are used for the measurement of**
   (a) Gauge pressure
   (b) Condensation temperature
   (c) Concentration of suspended materials in air
   (d) Humidity

9. **Which instrument is used for the measurement of dielectric loss**
   (a) Maxwell (b) Anderson
   (c) Schering Bridge (d) Kelvin

10. **Sensitivity of measurement is delines**
    (a) The maximum error
    (b) Change in the output with the change in the quantity being measured
    (c) Maximum error the instrument may result
    (d) Minimum measurement an instrument is capable of measuring

11. **Chromometer device is used for ....**
    (a) To measure the spectrum of lights emitted by metals when heated
    (b) To measure time
    (c) To measure distance
    (d) To measure very light weights

12. **Principle of Hydrometer ....**
    (a) Finding the volume of the water displaced when known weight is dipped
    (b) Measuring the level of measuring of weight of known volume
    (c) Vibration is sent and resonant frequency is found out which is dependent on
    (d) Measuring the level of floating in the liquid in reference to water the density of liquid

13. **What happens to the density of air with increase in humidity?**
    (a) Increases
    (b) Decreases
    (c) Density is linked to many other parameters and cannot be predicted
    (d) No effect

14. **Principle effect given by .... is used for the measurement of magnetic flux**
    (a) Hall (b) Piezo
    (c) Seeback (d) Kelvin

15. Precision of a measuring device aware about

(a) The degree of freedom from random error

(b) The closeness of the readings when the same input is applied repetitively

(c) How close the output reading of the instrument is to the correct value over a short period of time

(d) All of these

16. Hygrometer device is used for the measurement of

(a) Temperature (b) Humidity in air

(c) Pressure (d) Sound Level

17. LVDT stands for

(a) Linear Variable Displacement Transformer

(b) Linear Variable Differential Transformer

(c) Linear Variable Displacement Transducer

(d) All of these

18. What 122°F will read on a Centigrade Scale?

(a) 162 (b) 54

(c) 50 (d) 44

19. Which is not an electronic hygrometer

(a) Capacitive (b) Thermal Conductivity

(c) Resistive (d) Metal Paper coil Hygrometer

20. Variable head flow meters can be used for measurement of

(a) Liquids flow only (b) Liquids and gases flow

(c) Slurries flow only (d) Liquids gases and slurries

21. The working principle of Mercury Barometric Pressure is

(a) Balancing the air column with that of Hg

(b) Balancing the density of air with that of Hg

(c) Balancing the weight of the air column with that of Hg column

(d) All of these

22. Transducer's function is to

(a) Modify the input signal

(b) Codify/decodify the input signal

(c) Convert the primary signal into a more useful quantity, usually an electric impulse

(d) Amplify the input signal

23. **Rotameter's operation is based on**
(a) Pressure drop across a nozzle (b) Variable flow area
(c) Rotation of a turbine (d) Pressure at a stagnation point

24. **Positioning controllers are used for**
(a) Low loads (b) High loads
(c) Temperature changes (d) Flow rate changes

25. **Load cells are used for the measurement of**
(a) Weight (b) Strain
(c) Stress (d) Velocity

26. **Which of the following is the dynamic characteristics of an instrument?**
(a) Sensitivity (b) Reproducibility
(c) Dead zone (d) Fidelity

27. **............. is a functioning element of vapor pressure thermometer**
(a) Pointer (b) Bulb
(c) Bourdon tube (d) None of these

28. **Which of the following vapor pressure thermometer is a secondary element**
(a) Pointer (b) Bulb
(c) Bourdon tube (d) None of these

29. **PERT stands for**
(a) Program Evaluation Review Technique
(b) Process Evaluation Review Technique
(c) Program Equivalent Review Technique
(d) Process Equivalent Review Technique

30. **The temperature of tempering oil baths maintained at 400°C at the time of heat treatment of steel is measured by a/an ....thermocouple.**
(a) Iron-constantan (b) Chromel-alumel
(c) Platinum-platinum/rhodium (d) None of these

31. **Smoke density of the flue gas going out of the chimney is measured by a**
(a) Chromatograph (b) Polarograph
(c) Thermal conductivity meter (d) Photo electric cell

32. **Contact pyrometer...............**
(a) Optical pyrometer (b) Infra-red pyrometer
(c) Resistance pyrometer (d) Radiation pyrometer

33. The amplitude ratio for the sinusoidal response of ....is < 1

(a) Transportation lag
(b) First order system
(c) Second order system
(d) None of these

34. Starting temperature of optical radiation pyrometer is

(a) 1500
(b) 1200
(c) 800
(d) 400

35. Pirani gauge device is used for the measurement of

(a) High vacuum
(b) Very high pressure
(c) Liquid level at atmospheric pressure
(d) Liquid level under pressure.

36. Moving iron ammeters are used for measurement of

(a) AC only
(b) DC only
(c) DC as well AC
(d) AC, DC, and Watt

37. Unit of luminous intensity is ....

(a) Lux
(b) Lumen
(c) Candela
(d) Lumen/watt

38. What is the reason of pressure reduction at higher altitude

(a) Reduced height of air column above
(b) Air hits the mountain and return backs to plain thus reducing air density
(c) Due to cold weather at mountain, the air is having a tendency to move out to plains
(d) None of these

39. Which technique is followed for the measurement of very low value of vacuum of the order of 10

(a) Thermocouple gauge
(b) Ionization vacuum gauge
(c) Pirani gauge
(d) Compound Pressure gauge

40. ....controllers has the least maximum deviation

(a) P-D controller
(b) P-I controller
(c) P-controller
(d) P-I-D controller

41. Which of the following is undesirable in thermocouples used in industries

(a) Oxidation resistance

(b) Non-linear relation of emf to temperature

(c) Linear relation of emf to temperature

(d) Corrosion resistance

42. Which of the following is not a differential pressure flow meter?

(a) Flow nozzle (b) Rotameter

(c) Venturimeter (d) Orifice meter

43. The maximum service temperature for fiber glass used as thermocouple wire insulation material is .... °C

(a) 750 (b) 500

(c) 1000 (d) 250

44. The temperature range for copper resistance thermometer can be used is...................... °C

(a) 350 to 750 (b) 0 to 150

(c) -200 to 150 (d) 500 to 900

45. Rotameter can be used

(a) Only in horizontal pipe (b) Only in a vertical pipe

(c) Only for air (d) In a horizontal pipe line with modifications

46. An amplitude ratio of 0.1 corresponds to .... decibels.

(a) -20 (b) -10

(c) 20 (d) 10

47. Hot wire anemometer is used for ....

(a) Gas velocities (b) Temperature

(c) Discharge (d) Pressure

48. A constant volume gas thermometer employing .... to measure sub-zero (i.e., < 0° C) temperature.

(a) Nitrogen (b) Hydrogen

(c) Helium (d) None of these

49. The deflection of the free end of the bimetallic strips in a bimetallic

(a) Linear (b) Non-linear

(c) Parabolic (d) Hyperbolic

50. Which of the following device cannot measure liquid discharge from a tank or reservoir

(a) Notches (b) Weirs

(c) Mouthpieces (d) Orifice meters

51. The sensitivity of an orifice meters

(a) Less than Venturi meter (b) Greater than more flow meter

(c) Greater than venture meter (d) None of above

52. Which of the following flow meter is independent of fluid density

(a) Electromagnetic flow meter (b) Turbine flow meter

(c) Ultrasonic flow meter (d) Laser doppler anemometer

53. Flow totalization is measured by

(a) Rota meter (b) Orifice plate

(c) Ultrasonic flow meter (d) Turbine flow meter

54. .... device is used to give temperature indication

(a) Thermistor (b) Transistor

(c) Thermocouple (d) LED

55. Which property or quarta crystal changes with the change in temperature

(a) Capacitance (b) Inductance

(c) Resistance (d) Resonant frequency

56. Which material is used for making the sensing element of the industrial pressure thermometer?

(a) Brass (b) Steel

(c) Both a and b (d) Constantan

57. In a pH meter, the function of reference electrode is

(a) Produce a constant voltage (b) Measure average pH value

(c) Produce a constant current (d) Provide temperature compensation

58. ALVDT has

(a) One primary coil and two secondary coils

(b) Two primary coil and one secondary coil

(c) One primary coil and one secondary coil

(d) Two primary coil and two secondary coils

59. For which type of measurement is used in capacitive transducer

(a) Transient measurement (b) Static measurements

(c) Both a and b (d) Dynamic measurements

60. **Which of the following flow meters is capable of representing the rate of flow as well as the totalization flow?**
(a) Nutating dise flow meter (b) Electromagnetic flow meter
(c) Orifice meter (d) Lobbed impeller flow meter

61. **The head loss of an orifice meter is**
(a) Less than that of the venturimeter
(b) Less than that of the nozzle flow meter
(c) Greater than that of the flow meter
(d) Greater than that of the venturimeter

62. **Electromagnetic flow meter are used to measure the**
(a) Flow of non-conducting fluids
(b) Flow of non-conducting fluids in a metallic pipe
(c) Flow of conducting fluids in a plastic pipe
(d) Wind velocity

63. **In an open channel, which flow meter is used for measuring the flow rate**
(a) Orifice meter (b) Ultrasonic flow meter
(c) Rotameter (d) Weir

64. **What is the order of minimum displacement that can be measured in .... capacitive transducers**
(a) 1 μm (b) 1 mm
(c) 1cm (d) 1x10-12m

65. **A photo conductive cell is useful for**
(a) High frequency applications (b) Medium frequency applications
(c) Low frequency applications (d) All of the above

66. **The dynamic characteristics of capacitive transducers are similar to those**
(a) High pass filters (b) Notch filters
(c) Band stop filters (d) Low pass filters

67. **A hall effect transducer can be used for measurement of**
(a) Current (b) Power
(c) Displacement (d) All of the above

68. **Rotameter is used for**
(a) Drag force flow meter (b) Rotating propeller type flow meter
(c) Variable area flow meter (d) Variable head flow meter

69. Which of the following is positive displacement device

(a) Sliding vane flow meter (b) Rotameter

(c) Volumetric tank method (d) Turbine flow meter

70. The tube of the rotameter is

(a) Tapers upwards (b) Tapers downwards

(c) Is of uniform cross section (d) Tapers in horizontal direction

71. Which of the following device is used for clean fluids only

(a) Turbine flow meter (b) Hot wire anemometer

(c) Ultrasonic flow meter (d) Laser doppler anemometer

72. …………………….. optical transducers is an active transducer

(a) Photo emissive cell (b) Photo transistor

(c) Photo diode (d) Photo voltaic cell

73. A tachometer encoder has …………………..output

(a) One (b) Two

(c) Three (d) All of above

74. A linear displacement transducer (digital) normally uses

(a) Straight binary code (b) Gray code

(c) BCD (d) Hexadecimal code

75. Incremental counters use

(a) One channel

(b) Two channels

(c) Two channels and sometimes three

(d) None of above

76. LEDs emit light

(a) Only in red color (b) Only in yellow color

(c) Only in green color (d) In red, green, yellow and amber color

77. In barometers the error caused by mercury vapour pressure is of the order of

(a) I kN/m (b) 1 N/m^2

(c) 0.1 N/m (d) 10 N/m

78. Inductive pressure transducers

(a) Are sensitive to vibrations (b) Have high sensitivity

(c) Poor dynamic response (d) None of the above

79. The desirable property of monometric fluid is

(a) High viscosity

(b) High co-efficient of thermal expansion

(c) Low vapour pressure

(d) Corrosiveness and stickiness

80. Pressure transducers using LVDTS have

(a) Poor dynamic response (b) Linear response over a large range

(c) Low sensitivity (d) All of the above

81. A typical piezoelectric pressure transducers has .... range

(a) 0-10 kPa (b) 1-100 MPa

(c) 0-700 kPa (d) 0-20 MPa

82. A potentiometer type accelerator can be used for measurement of vibration .... of frequencies

(a) Lower than 50Hz (b) Higher than 1kHz

(c) Higher than 10kHz (d) From 100Hz and higher

83. A buffer amplifier gain ....

(a) Infinity (b) Zero

(c) Unity (d) Dependent upon the circuit parameters

84. Hydraulic load cells are available which have a maximum capacity of

(a) 10 KN (b) 50 KN

(c) 10 MN (d) 50 MN

Answer Key

1.	(a)	21.	(c)	41.	(b)	61.	(c)	81.	(d)
2.	(b)	22.	(c)	42.	(b)	62.	(c)	82.	(a)
3.	(d)	23.	(b)	43.	(b)	63.	(d)	83.	(c)
4.	(d)	24.	(b)	44.	(c)	64.	(a)	84.	(d)
5.	(d)	25.	(a)	45.	(d)	65.	(c)		
6.	(a)	26.	(d)	46.	(a)	66.	(a)		
7.	(a)	27.	(a)	47.	(a)	67.	(d)		
8.	(a)	28.	(c)	48.	(c)	68.	(c)		
9.	(c)	29.	(a)	49.	(a)	69.	(a)		
10.	(b)	30.	(a)	50.	(d)	70.	(b)		
11.	(d)	31.	(d)	51.	(d)	71.	(b)		
12.	(c)	32.	(b)	52.	(a)	72.	(d)		
13.	(b)	33.	(b)	53.	(d)	73.	(a)		
14.	(a)	34.	(c)	54.	(c)	74.	(b)		
15.	(a)	35.	(a)	55.	(d)	75.	(c)		
16.	(b)	36.	(c)	56.	(b)	76.	(d)		
17.	(b)	37.	(c)	57.	(a)	77.	(b)		
18.	(c)	38.	(a)	58.	(c)	78.	(b)		
19.	(d)	39.	(c)	59.	(d)	79.	(c)		
20.	(d)	40.	(a)	60.	(d)	80.	(a)		

Matching

1.	Plant site	A	Dairy floors
2.	Inventory	B	E.O.Q. model
3.	Plant waste	C	Raw Material
4.	Sanitation	D	Disposal
Ans. 1-C, 2-B, 3-D,4-A			

1.	Product flow rate	A	Multi storey building
2.	Independent demand inventory management	B	Valuable management tool
3.	Roller floor mill	C	E.O.Q.
4.	Inventory model	D	Process planning
Ans. 1-D,2-B,3-A,4-C			

1.	Physical verification	A	Detergent
2.	Wetting agent	B	Inventory store
3.	Anaerobic digester	C	Advertisement
4.	Sales promotion	D	Waste treatment
Ans. 1-B,2-A,3-D,4-C			

1.	Straight line method	A	Inventory control
2.	Flow pattern	B	Ventilation
3.	Exhaust fan	C	L-type
4.	ABC analysis	D	Depreciation
Ans. 1-D,2-C,3-B,4-A			

1.	Plant Layout	A	Disposition of various facilities
2.	Plant	B	Manufacturing products
3.	Process layout	C	Factory act, 1948
4.	Law connected with working condition in factories	D	Functional layout
Ans. 1-B,2-A,3-D,4-C			

1.	Wages related law	A	Maternity benefit act, 1961
2.	Worker association related law	B	Trade union act, 1926
3.	Social insurance related law	C	Inventory control
4.	Function of material management	D	Workman compensation act, 1923
Ans.1-D,2-B,3-A,4-C			

1.	Sales	A	Advertising
2.	Accounting	B	Art & science of business transactions
3.	P- chart	C	Fraction defective
4.	Economic order quantity	D	Inventory carrying cost
Ans. 1-B,2-A,3-C,4-D			

1.	Industrial operation	A	Circular flow
2.	Short term forecasting	B	Vogels approximate method
3.	Transportation problem	C	Period of 5,10 & 20 year
4.	Flow pattern	D	Time study
Ans. 1-D,2-C,3-B,4-A			

1.	Foundation of scientific management	A	Normal time + allowances
2.	Founder of motion study	B	F.W. Taylor
3.	Economic order quantity	C	Frank Gilbreth
4.	Standard time	D	Square root of 2AP/C
Ans. 1-B,2-C,3-D,4-A			

1.	Purchasing	A	Sinking fund technique
2.	Labor efficiency	B	Time and motion studies
3.	Demineralization	C	Water for utilities
4.	Depreciation	D	Acquisition of goods in exchange of funds
Ans. 1-D,2-B,3-C,4-A			

1.	Thermal diffusivity	A	Double toned milk
2.	Non-Newtonian fluids	B	Thermal properties
3.	Ice Cream	C	Over –run
4	Milk fat < 2%	D	Rheology
Ans. 1-B,2-D,3-C,4-A			

1.	Surface heat exchanger	A	Vacreation
2.	Plant sanitation	B	UHT processing
3.	Aseptic filling	C	CIP
4.	Heat Treatment of cream under vaccum	D	Cooling of milk
Ans. 1-D,2-C,3-B,4-A			

1.	Hydraulic press	A	Clostridium botulinum
2.	Anaerobic micro-organism	B	Juice extraction
3.	α-lactalubumin	C	Microbial death rate
4.	Activation energy	D	Milk protein
Ans. 1-B,2-A,3-D,4-C			

1.	Instantization	A	Agitation
2.	Flow diversion valve	B	Fruits & vegetables
3.	Cabinet drying	C	Liquid food
4.	Drum drying	D	HTST
Ans. 1-A,2-D,3-B,4-C			

1.	Water purification	A	Chlorine
2.	Liquid droplet	B	Spray drying
3.	Water in oil emulsion	C	Butter
4.	Liquid food	D	Drum drying
Ans. 1-A, 2-B,3-C,4-D			

1.	Power number	A	Fick's law
2.	Heat exchanger effectiveness	B	NTU
3.	Molecular diffusion	C	Agitation
4.	Activation energy	D	Sterilization
Ans. 1-C,2-B,3-A,4-D			

1.	Coefficient of performance	A	Bacterial removal
2.	Blanching	B	Pasteurization
3.	Hill test	C	Fruits & vegetables
4.	Bactofugation	D	Refrigeration
Ans. 1-D,2-C,3-B,4-A			

1.	HTST	A	Pasteurization
2.	Dehydration	B	Reduction of moisture to bone dry level
3.	Chilling milk	C	280 K
4.	Diffusion of molecules	D	Concentration gradient
Ans. 1-A,2-B,3-C,4-D			

1.	Sugar preservation of food	A	Reynolds number
2.	Low efficiency of spray dryer	B	Water activity
3.	Food spoilage	C	High exit air temperature
4.	Friction factor flow through pipe	D	Microorganism
Ans. 1-D,2-C,3-B,4-A			

1.	Continuity equation	A	Re<2100
2.	Bernoulli equation	B	Re>2100
3.	Streamline flow	C	Law of conservation of mass
4.	Turbulent flow	D	Law of conservation of energy
Ans. 1-C,2-D,3-A,4-B			

1.	Liquid-liquid separation	A	Cream-separated milk
2.	Centralized system	B	CIP circuits
3.	Flow through filter cake	C	70-90%
4.	Efficiency of regenerator in pasteurizer	D	Pressure drop is varied
Ans. 1-A,2-B,3-D,4-C			

1.	Mould growth ceases at water activity	A	M=σα
2.	Micro porous food matrix	B	< 20°A
3.	Maximum binding energy	C	< 0.6
4.	Linear sorption isotherm	D	Zone-I
Ans. 1-C,2-B,3-D,4-A			

1.	Food freezing	A	Freeze drying
2.	Sublimation	B	Destruction of all micro organisms
3.	Homogenization	C	Equalization of fat particles
4.	Aseptic processing	D	Eutectic temperature
Ans. 1-D,2-A,3-B,4-C			

1.	Convective flow	A	BHA
2.	Antioxidant	B	Molecular size and solubility in polymer
3.	Activated diffusion	C	Molecular size and viscosity
4.	Type-II isotherm	D	Sigmoid
Ans. 1-C,2-A,3-B,4-D			

1.	Canning	A	Mushroom
2.	EMC measurements	B	Z-value
3.	Thermal resistance constant	C	Plate freezer
4.	Indirect contact system	D	Peas
Ans. 1-A,2-D,3-B,4-C			

1.	Emulsifier	A	D-value
2.	Decimal reduction rate	B	Milk
3.	Ascorbic acid	C	Reynolds number
4.	Ratio of inertia to viscous force	D	Anti-oxidant
Ans. 1-B,2-A,3-D,4-C			

1.	C. botulinum	A	Sterilization
2.	Poise	B	Freezing
3.	Psychrometer	C	Viscosity
4.	Plank's law	D	Relative humidity
Ans. 1-A,2-C,3-D,4-B			

1.	HTST pasteurization	A	Plate heat exchanger
2.	Liquid nitrogen freezer	B	Extrusion
3.	Single and twin screw	C	Triple point
4.	Lypohilzation	D	Cryogenic freezing
Ans. 1-A,2-D,3-B,4-C			

1.	Multiple effect evaporator	A	Steam economy
2.	Atomization	B	Ultra pasteurization
3.	Uperization	C	Sterilization
4.	Retorts	D	Spray drying
Ans. 1-A,2-D,3-B,4-C			

1.	Mixing	A	Pasteurizer with steam injection
2.	Ultra pasteurization	B	150 °C for <1 sec.
3.	Vacuum chamber	C	Power number
4.	Holding method of pasteurization	D	62.8-65.6°C for 30 minutes
Ans. 1-C,2-B,3-A,4-D			

References

Earle, R. L. (2013). Unit operations in food processing. Elsevier.

Fellows, P. J. (2000). Food Processing Technology: Principles and Practice, Wood Head Publishing. Cambridge, UK.

ICAR. (2012). Handbook of Agricultural Engineering, New Delhi.

Kudra, T., & Raghavan, G. S. V. (1991). Post-harvest technology of cereals pulses and oilseeds: oxford IBH publishing co. Pvt ltd. New Delhi, 1988. Drying Technology, 9(2), 527-528.

Michael, A. M., Michael, A. M., & Ojha, T. P. (1984). Principles of Agricultural Engineering Vol. I. Jain Brothers..

Mohsenin, N. N. (2020). Physical properties of plant and animal materials: v. 1: physical characteristics and mechanical properties. Routledge.

Robertson, G. L. (2005). Food packaging: principles and practice. CRC press.

Sahay, K. M., & Singh, K. K. (1996). Unit operations of agricultural processing. Vikas Publishing House Pvt. Ltd..

Sanjay, K. (2010). A numerical approach in agricultural engineering with objective, NET, GATE, and ARS Digest. Kalyani Publishers, New Delhi, 110002, 247-256.

Smith, P. G., & Smith, P. G. (2011). An introduction to food process engineering (pp. 1-3). Springer US.

Toledo, R. T., Singh, R. K., & Kong, F. (2007). Fundamentals of food process engineering (Vol. 297). New York: Springer.

Traitler, H., Coleman, B., & Hofmann, K. (2014). Food industry design, technology and innovation. John Wiley & Sons.

Tufail, A., & Tufail, A. (1990). Dairy Plant Engineering and Management. Kitab Mahal Agencies.

9 789359 196565

Printed at Repro India Ltd.